The Traditional Farming Year

The Traditional
Farming
Year

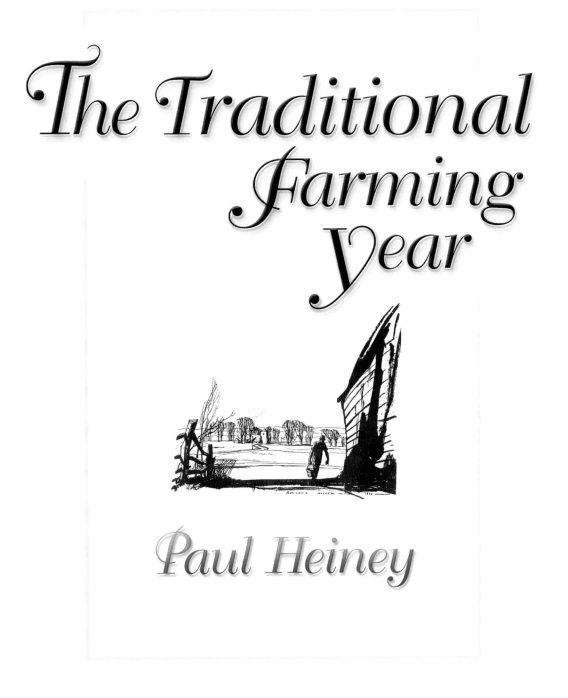

Paul Heiney

Old Pond
PUBLISHING

Published by

Old Pond Publishing
Dencora Business Centre, 36 White House Road, Ipswich, IP1 5LT
www.oldpond.com

Cover design and book layout by Liz Whatling
Printed and bound in Great Britain by Butler and Tanner Ltd, Frome and London

Contents

Acknowledgements

Chapter Heading Illustrations:
Caroline Church

Photographs:
Jonathan Brown and Caroline Benson at the
Museum of English Rural Life, University of Reading.
All photographs reproduced by courtesy of the University of Reading.

Front Cover Painting:
Ipswich Borough Council Museums and Galleries
for the use of detail from *Higham from
Langham* by John Millar Watt, 1927, oil on canvas.

Title Page and Half-Title Illustrations:
Rowland Hilder, courtesy of the estate of Rowland Hilder.

Text Line Drawings:
Reproduced from *Stephens' Book of the Farm* (5th Edition, 1908)
and *Farm Engineering* by John Scott (1885).
Further drawings by Felicity Halston.

The Traditional Farming Year

'Traditional' is a dangerous word. It means too many different things to many people. 'That's not the way to use a hay-knife ...' someone will shout, '... the way my grandfather did it, now *that's* the proper way.' Then another will mutter, 'That's not the way *my* old man did it. Now, his was the *real* traditional way.' No one who has lived any kind of a rural life can escape these arguments. The trouble is, traditions are being born all the time, in all walks of life: I imagine that someone soon will come out with a phrase such as 'a *traditional* way to use a mobile phone!' You can't point your finger at any date in history and say, with confidence, that things were traditional, then.

But I suspect that you and I know what we mean when we talk of a 'traditional farm'. The words conjure up some kind of image, a picture so real that we can almost reach out and touch it. If we had to try and describe a traditional farm, we might say it is simply a farm where the work of growing the crops and husbanding the animals is done according to rules which were not invented to meet the needs of modern science or economics, but which came out of the handed-down instincts and practices of generations of farmers that have gone before, and which worked well for them. There can never be anything new or cutting edge about traditional farming; it is by definition something that has emerged from the past, a common understanding that survives.

Before we try and work out what we mean by traditional, is there any such thing as a typical farm? Aren't there as many farms as there are farmers? If you walk the fells and hills of the north-west, you will find very little farming there that has anything to do with the working lives of those who grow corn in eastern England. The fruit growers of Kent, or the vegetable farmers of the Fens, live in a world far removed from that of the crofters of the Hebrides. Yet we bundle all these patches of land together under the heading of 'farms' and somehow think the same rules apply to all of them. Add to that

A traditional farmer returned natural fertility to the soil with muck or compost:
James Steels's Manor Farm, Grazeley, Berks.

confusion the uncertain word 'traditional', and you'll see why to describe this book as an account of a traditional farming year is a dangerous business.

So let me tell you what I believe traditional farming to be. First of all, it must be a system of farming in which nature remains in charge. I believe there is an unwritten agreement between the farmer and the natural forces that surround him. A deal has been struck between man and nature which offers the farmer a living in return for a little respect for the land and his animals. A decent traditional farmer, for example, would never think of growing a hungry crop, such as potatoes, without returning natural fertility to the soil with muck or compost. In return for his crop, the farmer gave back a little of what he'd taken from the land. That was the deal, and if he broke it he'd pay the price the following season with a poor crop. He didn't force-feed animals, or imprison them in buildings too small for their own good, because he knew

Poultry on stubble.

that such practices only made them sick, and ailing animals would never deliver anything. Instead, he fed them what they'd probably choose to eat if they were able to fend for themselves. His return was a supply of eggs, bacon, beef or lamb, wool and leather. None of these rules were written down. Farmers learnt over generations the principles of husbandry; they gathered knowledge of what worked and didn't work, and it was on this basis that they farmed their land.

If we now fast-forward to those farms where tradition has been swept aside, you will find less reliance on wisdom. The needs of the farmer have changed. In the affluent world in which we live, a farmer rightly asks why should he be the only one left leading a peasant lifestyle? Agricultural science has given him his escape from the bleakness of the old days and the old ways. Problems which confounded the traditional farmer for years have been solved by science over the last fifty. Any crop disease you care to mention now has its medicine, every predator has a spray to combat it. Pigs can be made to grow so that every pig is alike to meet the customers' thoughtless needs. There's no need for a hen to see the light of day in order to lay a profitable egg. The machines grow bigger; the yields from fields expand to meet them. And the first thing that gets left behind in the headlong rush forward, is tradition. Nothing that grandfather said has any importance any longer, because there's nothing left on the farm that the poor old devil would recognize. Little of what has been left behind matters much, only that which lies ahead. Tradition is of no value, only progress is worth chasing.

Which is not to say that tradition was always right, had all the answers, or provided every solution. There's much about it that is best dead and buried, such as pitiful rural poverty, and the exploitation of farm labourers. The old ways of doing things was a hungry and demanding beast to feed which led to crippling toil and premature death for many. But these are social, not agricultural problems. When it comes to growing food and caring for the land, none of this means that at the heart of traditional farming there isn't something worth preserving.

A scene from the 1930s on the Suffolk farm of Lady Balfour, one of the pioneers of the organic farming movement.

There is nothing in the past from which a lesson can't be learnt, which is why I believe that what we now think of as the 'traditional' way of farming is worth re-examining. Many of its practices were still familiar in this country as recently as fifty years ago, the best of them taken up by the new wave of organic farming. But the advancing tide of modernization has been so swift that already much of it reads like the practices of several centuries past. So, read this book for merely nostalgic reasons, if you wish. But give a passing thought also to the notion that although it may appear inefficient, out-dated, labour-intensive stuff, at the heart of it there might be a grain of truth, an agricultural wisdom to sustain us if we ever wake up to the fact that the way we farm today may not be all it's cracked up to be.   Grandfather may not have been as clever as he would have you believe, and he may not have been right all the time, but he was a damned good farmer. He *must* have been. He made a living, raised a family, kept his farm going with far less machinery and science than his successors have. So there must be something in that old, traditional farming, mustn't there?

The Farm

The traditional farming year was as dramatic a creation as it is possible to imagine. It has all the highs and lows of the most intense theatre. There are unremitting ups and downs in the fortunes of the people who played the leading roles - the farmers, horsemen, stockmen, farmers' wives, all the way down to the sow in the backyard sty. On every farm you would have found tales of great good fortune and financial catastrophe, of love and heartbreak, battles won and lost with the forces of nature. Not even the most florid of grand operas could do justice to the emotional roller-coaster of traditional farming life.

We think of the turmoil of the industrial revolution, and the pain of its progress, of the pioneering work of the early engineers digging canals, throwing bridges. But for sheer drama, farming can outmatch them all. To watch the farming year roll by is to be on the edge of your seat for each of its 365 days.

It is worth spending a little while to look at those who played their parts, who took centre stage at various times of the farming year, never performing from a script because, taken as drama, the farming year was entirely improvised with none of the players being certain what would happen next. Without doubt, the farmer himself was the star of the show, although not necessarily the hero. Farmers, then, were born and not made. It was unlikely that he had come fresh to the land, more probably he was following in his father's footsteps. Whether he owned his land, or was a tenant, made not much difference, other than to his economics. He knew that farming was a struggle, and to win the war needed troops on high and constant alert.

For this reason, every farmer worth the name had eyes in the back of his head. Nothing happened on the farm without him knowing about it, filing the details in his mind for future reference. It often made him a nagging presence, as his workers would tell you, but only by attention to every detail could a proper farm be run. When he gave orders, he expected them to be carried out, and to his standard. There was no discussion, no negotiation and no suggestion that the farmer was anything other than right, all the time. He was every bit as much the authoritarian as the mill-owner or army officer. An old horseman recounted to me the habit of a farmer for whom he once worked: 'When you'd a' finished harrowin' a field, he'd come and look at it, and if he found so much as a horse's footprint anywhere on it, he'd send you back to harrow the whole damned field again!'

Feeding stock on a bright winter's morning.

The kind of farmer you were was often dictated by the size of the farm itself. On a small, family farm, of less than a hundred acres, the farmer would be less a managerial figure, and more hands-on. In fact, he and his family might be the only pairs of hands on the entire holding. The family farmer came at the bottom of the farming hierarchy, did not employ an army of men, had no manor house and held no social rank such as an estate-owning farmer might. He did not employ a horseman, because he did it himself. Likewise, he was shepherd and stockman rolled into one. His wife cared not just for the family, but ran the dairy too, and chased hens for eggs. At lambing time, she nursed the sickly newborns by the stove. She too was far removed from the mistresses of the 'grand houses'; she had no servants but was a constant servant to others. Their children had few choices in life. Sons would follow in fathers' footsteps and from a very early age they learned to plough, sow, reap and mow. Daughters too learned their mothers' supportive crafts, and eventually married other

Marlborough 1944: a fine Jersey cow. In the background the haystack neatly sliced and plenty remaining for feed.

farmers' sons, for who else are they ever likely to meet? It was uncommon to leave a traditional farming community: to escape from the land, preferring a broader, richer life in the towns and cities, was to be some kind of traitor to one's roots. To be a farmer required your feet to be as firmly planted in the soil as the roots of your crops, and you did not wander far. In fact, the more traditional the farm, the less change ever took place. Jobs taken in teenage years were often held until retirement. The parish boundary was seemingly as impregnable as the perimeter fence of a prison; either that or there was little desire to see what life might be like on the other side of the hill. Farmers are rarely blessed with a wider vision, or learn to suppress it. The farming life could be summed up quite neatly in the traditional verse:

> *Man, to the plough*
> *Wife, to the cow*
> *Girl, to the yarn*
> *Boy, to the barn*
> *And your rent will be netted.*

The farmer may have been the boss on his farm, but the horseman came pretty damned close. To describe him as merely someone who cared for and worked the cart-horses, is to do scant justice to the job that carried with it an almost aristocratic status. And rightly too, for cart horses were the motive power of traditional farming, and without them hardly a job on the farm could be done. The man who had control of the horses, in truth had total charge of the farm. His skill, although hard learnt, was essentially a simple one: he had to achieve not only a trust with his horses, and a power over them, but a communication between man and horse which at times bordered on magical. It is not fanciful to suggest that true horsemen have special powers over their horses. Of course, there are exaggerated tales of just what they could achieve: stories are told of how a horseman could merely whisper a secret word into a horse's ear, and from that moment on have total control over it. To have knowledge of this magical word did not come easily either: in parts of the country, secretive horsemen's societies existed with their own humiliating initiation rights which had to be endured before the Word could be passed on to them. In Scotland, the initiation culminated in the young horseman 'shaking hands with the devil' which was, in fact, a stick wrapped in hessian sacking although the blindfolded youth didn't know this.

But once in possession of this magical word tradition has it that a horseman's powers were limitless. He could break a horse to harness, halt it in its tracks making it as motionless as if it were a statue, if he wanted. He would also be given the secrets of the tinctures and lotions which were yet another branch of the horseman's secret armoury. These mixtures, carried in small bottles, were cures for all ills in horses, both mental and physical. Doubtless, some of them worked; but the real object of the intense mystery that surrounded the working of horses was an understandable desire of a horseman to protect his job. If an entire community could be made to believe that a man was possessed of special powers over a cart-horse, and that he was forbidden to tell what those powers were, didn't he have a job for life? Of course, many horsemen did display such powers, although they more than likely came from an instinctive understanding and empathy with horses that came from a lifetime in their company, rather than from the little brown bottles, whispered words or secret ceremonies. However he achieved it, though, the horseman was undeniably the major figure on the traditional arable farm.

No such mystery seems to have been attached to the stockman. His charges were the cows, the bulls and bullocks and his job was to get as much milk as could be had

The highly respected shepherd with his dog and leg crook.

from the cows, and get as much fat as fast as possible onto the bullocks. No secret societies for him, none of the status that came with being able to work a cart-horse. This contrasts with the shepherd who was a figure worthy of great respect, a loner by nature who was allowed to get on with his work largely undisturbed. Like the horsemen, he too carried secret recipes for tinctures and lotions for curing the many ills to which sheep were susceptible. Sheep are a minor part of farming life these days; mere suppliers of spring lamb. But the traditional farm saw the sheep as a true provider not only of meat, but of wool which, in its time, commanded a high price. It was for good reasons that sheep were referred to as 'the golden hoof', for in their ramblings across the land, guided by the ever vigilant shepherd, not only did their cloven feet cultivate and break down heavy soils, but compact sandy ones. At the same time, they spread their own muck to give vigour to crops that followed in later seasons. Given that the sheep was such a generous creature, even if susceptible to every ailment nature could conjure up, it is little wonder that the job of shepherd had some status attached to it.

Farm labourers had to be able to bear weighty sacks of corn.

The lowest of the ranks was the farm labourer, but he was not without his skills, or his pride. Although he would be allowed use of a horse, he would carry no authority in the stable and if he had differing views from the farmer on how the farm should be run, he would keep them to himself if he had any sense. He was, in effect, something of a pack-horse himself. He would think nothing of carrying weighty sacks of corn up steep stairs to a granary, working day after day in the baking sun, collecting sheaves of corn, hoeing roots crops, turning hay with forks. Like the constant moaning of the wind, he may well have grumbled, endlessly, about his burden, but he'd get on with it. His skills were manual ones, and although less impressive perhaps than some of the tricks the horsemen could perform, he had his secrets too. A good farm worker had his own tools, for example, and guarded them. It would be a brave lad who borrowed another man's hoe, or axe. All jobs fell to the farm worker; he would be expected to dig ditches, erect fences, repair gates, build walls, thatch cornstacks, repair machines. But even he was not the lowest rank on the traditional farm.

Preparing an Ayrshire heifer for showing: Castle Douglas in the 1950s.

At the very bottom of the heap sat the hapless farmer's boy. If ever there was a dogsbody on the farm, this was he. Rarely would he have been the farmer's son, who would have had a little more protection, but a lad brought in from the village to mop up the jobs that no one else had time to do. Much of his labouring was domestic, which was why in Suffolk he was known as the back'us, or 'back house' boy, and came under the direct command of the farmer's wife. He would be the first out of bed, often before dawn, to light the early morning fires. The cleaning of boots and shoes would be on his list of tasks; he would scour the garden for vegetables, wash up, clean the dairy, peel potatoes, chop wood, pluck poultry, take letters to the village, dig the garden, and without complaint was expected to be the butt of the horsemen's jokes. And all the time, he would be gazing at the stables or the stockyard, waiting for the day when he would be free of his domestic duties and get to work on the land, which seemed to him to be his destiny.

These, then, are the characters that play out the drama called traditional farming. Each has a defined role, as precise as if they were ranks in an army. And now, let the dramatic events of the traditional farming year unfold.

January

It is the farmer's boy, the earliest riser on the farm, who has noticed that the days are getting longer by a few minutes every day. This, though, is of little comfort to the farmer who remembers the wise saying that, 'as the days lengthen, the cold strengthen.' Although midwinter, technically, has passed, there remains a sense of much winter weather to come. On some farms, winter will hardly have started yet.

The farmer hardly knows what kind of January will suit him best. If it is mild, as it can often be, then the omens are poor for, 'If the grass grow in Janiveer, It grows worse for it all year.' On the other hand, 'He who would fill his pouch with groats, In Januair must sow his oats.' So does the farmer kneel in the village church and pray for a little midwinter warmth to plant his oat seed, or would he prefer a few dank, chilly weeks to hold the grass in check before it exhausts itself with premature growth? This conflict, and others like it, will be a recurring theme in the months to come: what is right for one thing on the farm rarely suits another. Farming seems to have been invented to provide anguish for those who try to make a living from the land.

The New Year begins, for some, in church, for the celebration of Plough Sunday which traditionally falls on the first Sunday after Epiphany. To the church is brought a plough, its breast gleaming and paint refreshed. Prayers are said for a bountiful harvest, healthy stock, a kind climate, and they are deeply meant. Every farmer, God-fearer or not, knows that there is no predicting the farming year ahead, each recognizing that it only takes a rampant disease, a tempest or a drought, and the fragile economy of the farm can be shattered. The service ends with the rousing words, 'God Speed the Plough'. Without doubt, the ploughs will be moving with hardly an idle moment in the remaining days of this midwinter month.

1950s Hertfordshire: a well-laid hedge 'worth a roll of barbed wire any day'.

In a corner of a far field, a wisp of smoke rises from the remnant of a bonfire. Here, the hedger has been at work, taking advantage of the lack of leaf on the hedgerows to prune and shape them, not for needless artistic reasons but to preserve them as stock-proof barriers. A well laid hedge is worth a roll of barbed wire any day. But to achieve a thickness which will not only prevent a cow or a horse from jumping it, while at the same time preventing a sheep from shuffling under it, calls for no small skill on the part of the hedger. He wears thick leather gloves to protect him from the thorns, and leather trousers too if he has any choice. Then, armed only with a maul - a kind of cudgel for driving stakes - and a bill-hook and long-handled slasher for cutting and trimming, he will start to remove the surplus growth leaving a hedge which looks like a shadow of its former self. To an untrained eye it looks like destructive work, but in truth it is deeply creative. It is precise work, for when he delivers the blow from the bill-hook which half-severs the trunk of the standing stem,

Laying hawthorn with a hand bill.

he knows that an inch too far and he will kill it. Equally, an inch too little and it will never bend to his will and allow him to weave it between the hazel rods which will keep it in place. He works alone, many long hours, eating his food by the fire which consumes the trimmings. He makes his way home at night knowing that if he has done his job properly, then the hedge will need no more attention for five years, at least.

As those worshippers, fresh from the blessing of their plough, prepare to start another farming year, few will be able to look over those newly laid hedges and not remark on the standard of the half-finished ploughing which they surround. In fact, the criticism of others' work was almost a rural sport in traditional farming days. A gang of horsemen might wander the parish, seeing how Jake had opened up his furrows, or how old Arthur had drawn out his. They were looking, literally, for any departure from the straight and narrow, and if they found any then the culprit would be severely mocked that night if he dared show his face in the pub. 'Ploughing! That looks more like a dog's pissed in the snow,' they'd bellow, and the hapless horseman would have to bear the ridicule till someone else was discovered doing less than perfect work, and the butt of the cruel joke would shift to him.

The working day started early whatever the season. There were no concessions to frost and snow. If the farm were large and prosperous enough to have the services of a farmer's boy, then he would be the first to rise at the cruelly early hour of five in the morning. Finding his way around the cold farmhouse with the help of a candle or an oil lamp, his first and most urgent task would be to get some warmth into the place. There was a time, which in parts lasted into the early twentieth century, when the central heating was provided by the livestock, and not the kitchen stove. Look at a Dutch farmhouse, for example, and you will see the living accommodation directly above the cow stalls where the livestock remained for the entire winter. Each cow was as good as a radiator, and the warmth from it spread upwards through the cracks between the oak floorboards, and warmed the family above. But on our farm, it is the boy's job to gather together dry kindling and encourage a blaze in the grate of the kitchen range on which he will place the large kettle to greet the farmer with early morning tea.

A sound in the farmyard disturbs him; it is the clatter of hobnailed boots on the cobbled yard and he guesses it is either the first of the horsemen arriving, or the stockman getting ready for milking time. Whichever it was, they were both intent on making life hell for the farmer's boy whose job it was to be on the receiving end of every demand, and the butt of every ounce of blame that flew around the farmhouse. Many young lads couldn't stick it, but for those who could the rewards were often worth the suffering. Some would become assistants to the shepherd, known as a 'page', or join the gang of stockmen. Some dreamt of joining the ranks of the horsemen who were the aristocrats of the farm. Few farmer's boys ever had thoughts of becoming landowners themselves: that was a dream too far.

The early morning scene is set, then, and it will vary little throughout the farming year. Only the time of sunrise casts it in a different light. If there are cows to be milked, then that will be the first, urgent job of the day, the milk drawn by hand into scalded pails provided sparkling and fresh every morning by the farmer's boy – one more job.

January is the height of the ploughing season, when horses are working at their hardest and the feeding of them was as vital a job on the traditional farm as the filling

Three plough teams with a large acreage to prepare.

of a tractor with fuel on its modern counterpart. Much will depend on the weather, but the horses might have spent the night in a yard, deep in straw and muck, the base of which would have been laid down in November. The rich and warming mixture of bedding and dung would have been allowed to thicken over the winter months. In mild winters, and especially dry ones, the horses could have spent the night in a near-by meadow which would have provided no feed, or course, but if the land was dry and the

Heavy going: winter ploughing with three horses.

Suffolk horses in their stalls at Worlingham, 1940.

horses' vast feet did not churn up the land, there was no harm in a cart-horse spending a winter's night in the open. Horses have a strange sense of cold: you might see one happily standing in the middle of a field in a blizzard when there was plenty of shelter to be had from a nearby hedge and wonder why they don't take advantage of it. This is why, on cold and damp nights, a careful horseman would prefer to see them in a stable, safe from the elements.

The farm stable often consisted of a series of stalls in which each horse would be tethered. At the head of each stall there would be a manger for his feed, and above it a rack for hay. A gangway ran the length of the stable, just behind where the horses stood, and this is where they would dung. On very cold nights, when the doors had been tightly shut and the windows closed, to enter the stable the next morning was to take your breath away as the heady smells of ammonia from the urine and the steaming dung, filled the air. Whichever horseman was the most junior took the job

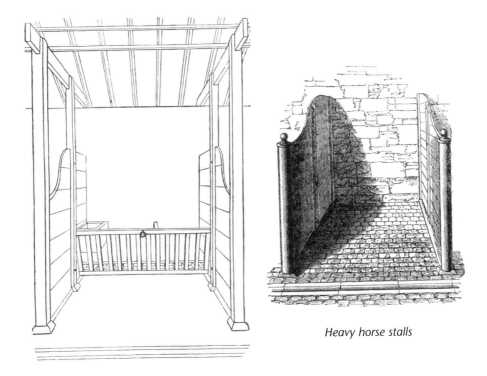

Heavy horse stalls

of mucking out, and took it seriously too. It was far from being a simple task of merely flinging the droppings aside, but required rather careful use of the pitchfork to separate dry straw from wet, preserve the former and get rid of the latter. And whilst that was being done, the horses would fidget, stamp their feet, swish their tails, perform any trick they could to draw the horseman's attention to the fact that their manger was empty and already there was a hint of daylight in the sky. Work was looming: they want fuel, and they want it now!

It could well be that on a large farm, the farmer himself might have no idea how his horses were being fed. A horse's rations were yet another of the horseman's little secrets which set him apart from the others and helped to secure his employment. Oats were the staple diet of a working horse, which were usually lightly rolled to break the husk, making the digestion of them easier. But a diet of oats and nothing else would become monotonous for a horse, and so to give variety the cereal was often mixed with chopped hay, or preferably oat straw, to provide some bulk. Recipes were numerous; bran was sometimes added to the cereal to give more roughage, and a rich sweet liquor was made from thick, black molasses which was diluted with water and added to the feed making it irresistible. But farm horsemanship was a competitive

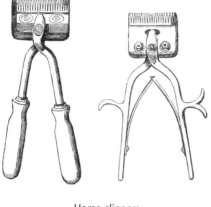

Horse clippers

business, and the feeding of the horses was as much to ensure that they were kept in better condition than the neighbour's, as it was to provide power for the plough. A good horseman wanted the best looking horses in the parish and might even stoop to a little harmless theft if that could provide the fodder he needed. For example, it was a well known fact that a little linseed in a horse's diet did wonders for the shine on its coat, and the nearest supply of this oil-rich food was in the form of the linseed cake which was fed to the milking cows. Some farmers, aware that a shiny horse went no faster than a dull coated one, viewed this as extravagance and would arrange for the cake store to be firmly locked and bolted, accessible only by the use of his key. Linseed cake was the only feedstuff which couldn't be grown on the farm, and had to be paid for with hard-earned cash. A juicy root would also do a horse no end of good, preferably in the shape of a mangel-wurzel, at which a horse would chomp happily for hours, sucking the sweetness from it. But the farmer grew mangels and turnips to feed sheep, cattle and pigs in the winter, not horses. Many an argument over fodder has rattled around the traditional farmyard as horseman and stockmen have hurled accusations of theft.

Horses would be allowed to eat in peace and quiet; for all its bulk and likeness to a tank, a cart-horse's digestion is a delicate affair. Within that huge frame lies a lengthy gut which is prone to kinking, like a garden hose. There was, and still is, no guaranteed cure for a horse which has twisted its gut, and any upset to a horse while feeding could therefore contribute to a disaster. Horses were left to feed in peace while horsemen fed themselves, and allowed a full hour for digestion. Only then, when the horse had licked the very last vestige of food from the dark depths of his manger, would the horseman return with a comb and brush to groom his coat till it shone like silk and the horse was ready to face the working day.

The day's labour often began before dawn, the ploughmen and their horses making their way down the lanes as much by instinct as by the aid of their eyes. On the shortest days, it would not be unusual for the men to be standing at the edge of the field, waiting for sufficient light in the sky for them to be able to see the distant end of the furrow. Only then could they begin their ploughing.

If there was any light to be seen in the field, it would probably belong to the shepherd, watchful over his ewes, alert for any early lambs. Although lambing was traditionally arranged for much later in the year, on lowland farms where the frosts were less severe and wet weather less likely, a farmer might risk some of his ewes lambing early in the year in order to fatten them quickly and get the very best prices at the spring markets. It was a gamble, of course, for there was no grass on which lambs could fatten, no pasture the ewe could graze to fortify her supply of milk. Instead, the shepherd fed the ewes mangels, cereals, turnips and hay, all of which came at no small cost. If the premium price of the resulting lambs failed to cover the increased costs, then it was one more potential disaster that could turn a farm from profit to loss. For that reason, the shepherd will be ever vigilant, ready to help any ewe in difficulty, quick to ensure the sodden newborn lamb is dried, taken to shelter, and gets the first, vital feed of its mother's milk. The shepherd will be no less proud

Fitting a dead lamb's skin over an orphan to persuade the bereaved ewe to accept it.

The shepherd's hut: wooden, with a corrugated tin roof, on an undercarriage.

of his flock than the horseman of his stable. In fact, he might have deserted his own home to live with the flock, full-time, till the difficult early lambing is over, for as long as six weeks. He is the only worker on the farm who can never entirely rest, for sheep are unpredictable, especially at lambing time, and he can never be far from them.

It is the orphan lambs which cause him the most work. These creatures have been born to mothers who, for one reason or another, have died and been unable to give their offspring that vital 'licking into shape' or offer the warm milk from their teats. Instead, the shepherd must provide the maternal comfort in those first days of life, bringing the lambs into his hut, placing them by the stove, persuading them that milk from a bottle is as good as their mother's. The shepherd's hut is a remarkably snug dwelling; wooden with a corrugated tin roof built on an undercarriage so a horse might pull it from field to field as the shepherd requires. For the lambing season it is

Inside a shepherd's hut in 1955. Walter Jacklin of Hackthorn, Lincs has electric light and a television.
He lived in his hut for six weeks a year.

as much home to him as his cottage. As the horsemen stood at the edge of the field waiting for the light, they might see a billow of smoke rising from the chimney of the hut as the shepherd stokes the fire to warm himself and the orphan lambs.

Farm cart

Ploughing cannot continue if the land is frozen, as it might well be on hard, January mornings. This does not mean that men or horses can remain idle. Instead of their plough harness, the horses will be wearing gear for shaft work and will be harnessed to the carts to shift the ever-growing mountain of muck which has come from stable, byre and pig pens. On a traditional mixed farm, muck is never in short supply when livestock have been brought in from the meadows and confined in yards for the winter. A cart-horse alone drops an average of 33lb of dung a day, and spills 12lb of urine. Multiply this by the cattle, sheep and pigs and over a typical winter a farm may gather a thousand tons of manure. As fertiliser, it is not at its best when fresh. It needs to compost, which requires it to be built into heaps and allowed to heat, being turned if necessary to ensure it rots evenly throughout - one more job for the manual labourer. Proper compost might take as long as six months to make, and then it will become a rich, friable, nourishing soil which the farmer can return to the land with confidence.

A farm worker is used to the smell of muck, and the job holds no fears for him. Indeed, the fierce chemical and biological reactions that are part of the composting process are happening at the very heart of the muck heap, turning it from waste into fertiliser, and giving off heat in huge quantities in the process, rising as steam on a frosty morning. The application of muck to the fields, which was a common practice in January, was no haphazard affair. Horse and tipping cart, often called a tumbril, would be driven to the heap, reversed till alongside it, and then filled forkful by forkful until the load was such that the horse needed to lean into his collar with a groan to get it on the move. If hill work was involved, one horse alone may be insufficient, and so a trace horse would be added directly ahead of the horse in the shafts, to give extra pull. The driving of a pair of horses, one in front of the other, called for rather more skill than merely leading one horse alone. It was possible to

Using a trace horse to give an extra pull for a cartload of muck.

drive them in such a way that the trace horse, in front, did all the work and allowed the horse in the shafts merely to jog along. Equally, if the lead horse wasn't kept 'up in its collar' then it might as well not have been there at all. Ensuring both horses shared the labour called for a horseman's skill, especially when negotiating a gateway or a tight bend, when it would be easy enough to end up with two horses pointing in opposite directions. It would not have been a popular horseman who caused a load to spill and the men to be brought from the yard to recover it, forkful by heavy forkful.

As January ends, it will have become noticeable that work in the fields can start that little bit earlier, and finish as much as half an hour later. But the days are the only things that show any sign of growth, with the possible exception of the young lambs. The haystack is dwindling, so is the heap of animal feed in the barn. The farmer casts an anxious eye over his reserves and hopes that soon he can turn the corner and put winter behind him.

February

Every farmer looks and hopes for a sign of change in February, and as early as Candlemas Day which falls on the second of the month, he will hold at the back of his mind the tried and tested thought that:

> *If Candlemas Day be fair and bright,*
> *Winter will have another flight;*
> *But if it be dark with clouds and rain,*
> *Winter is gone and will not come again.*

Also at the back of his mind will be the old reminder:

> *Lock in the barn on Candlemas Day*
> *Half your corn and half your hay.*

The first saying the farmer can choose to believe or ignore, but the second needs to be taken firmly to heart. During the course of this month he may well feel the fleeting warmth of the sun on his back, or a hint of balminess in the breeze, but however the fickle weather falls he is only halfway through the winter and there will be many weeks yet before livestock can feed themselves from the meadows. Man is fooled more easily than grass, and is often tricked into thinking the seasons have turned when, in fact, the meadows would be a more reliable guide to the progress of the year.

But the farmer knows that however much the seasons may tease him, they will come right in the end - they always have. The horses are often hard at plough this

———————— ◆ ————————

Plough teams at Eastbury, Berkshire.

month, re-working the fields which they last trod in the closing months of the previous year. The traditional farmer had few weapons in his armoury with which to deal with weeds, but the plough was certainly one of them and he knew that with its repeated use he may not completely win the war against weeds, but he could put up a damned good fight.

One of the major functions of ploughing, other than to provide a decent seedbed for the subsequent crop was to bury the growing weeds, smother them with soil, kill them off. It was, and still is, effective, but even the best ploughed field will soon be awash with weeds that sprout from the buried seeds brought freshly to the surface: once in the warmth and moisture they are only too happy to burst into growth. For that reason, the dedicated arable farmer will 'cross plough' his land. Having run the furrows one way up and down the field the previous autumn, in the spring he will run the plough the other way, across the field, cutting the original furrows at right angles,

Whippletrees

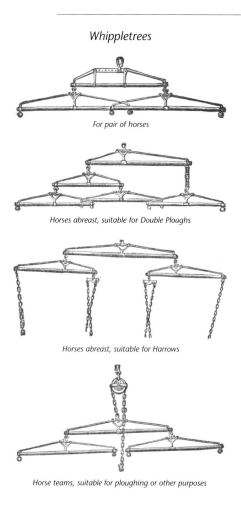

For pair of horses

Horses abreast, suitable for Double Ploughs

Horses abreast, suitable for Harrows

Horse teams, suitable for ploughing or other purposes

Gathered up ridges from the flat

beating into submission any weed that dares to think of bursting forth.

February can be a cold month but often a dry one, and this made it ideal for cross ploughing which would fail if the ground was too wet, for the weight of the horses' feet on the land would do more harm than good. It is a month for catching up, making sure you are on schedule to greet the spring with open arms when it finally arrives. Land must be made ready now, or it will never be ready at all.

Horses don't always find it easy to cross plough, not at first, anyway. They are creatures of huge intelligence, blessed with long memories. No matter which field they are taken to, they will remember the last time they were there, what they did and in which direction they were driven. But when cross-ploughing, the horsemen has to teach them new tricks for now the direction is across the field and not up and down it, and the first handful of furrows can be a struggle for both man and horses.

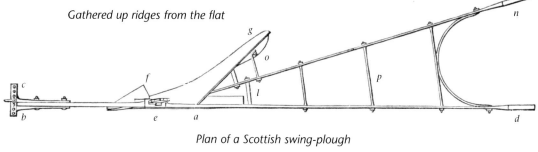

Plan of a Scottish swing-plough

b d Straight line of body.	a n Little stilt.	l Heel.
a b Beam.	e Head of coulter.	o Bolts fastening the
c Bridle.	f Feather of stock.	mouldboard to the little stilt.
a d Great stilt.	g Ear of mouldboard.	p Stays to support the stilts.

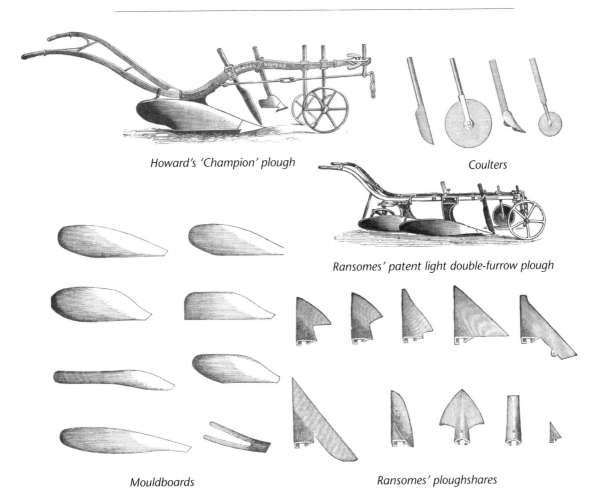

Howard's 'Champion' plough

Coulters

Ransomes' patent light double-furrow plough

Mouldboards

Ransomes' ploughshares

Anyone who has spent time with cart-horses, observing the way they bend to their task, the thoughtfulness they often apply to it, and their willingness, comes away with the hugest of respect for these animals - once the power house of agriculture. As the nineteenth century turned into the twentieth, and as some of greatest strides in agriculture were about to be taken, H. Rider Haggard wrote this in the diary of his own Norfolk farm of the cart-horse:

> *The intelligence evinced by farm horses at ploughing, and indeed all other work - if only you are master of the language which they understand - always strikes me as astonishing. The carriage and riding horse is generally very much of a fool and misbehaves himself, or gets frightened, or runs away upon most convenient occasions. How different it is with his humble farmyard cousin, who, through heat and cold, sun or snow, plods on hour after hour at his*

appointed task, never stepping aside or drawing a false line, always obedient to the voice of his driver, and, provided he is fairly fed and rested, always ready for his work the long year through. I often wonder whether, taken as a class, the common plough horse is really more intelligent than the aristocrat of the stable, or whether it is simply that the latter has, as a rule, so little to do and so much to eat that he seldom comes to understand the responsibilities of life.

However, not all cart-horses were angels, and those that did behave as extensions, if you like, of the men who had charge of them, did so not through a desire to please but as a result of hard lessons learnt at a young age. February was often a good month for foals to be born. There was a belief common amongst stockmen and horsemen that if new-borns were allowed to 'grow with the year' then they were healthier for it. It makes some kind of sense. Young stock born early in the year has the company of its mother for a few months before being weaned, by which time it is well into spring, and instead of the closeness of its mother it now has the sun for warmth and comfort. Also, as it learns there are other things in the world to feed off than its mother's milk, the grass on the meadows is becoming sweet, nutritious and tempting. As the year moves on, the foal or calf has learnt the trick of survival and is in good shape to face its first winter.

Foals did not learn the disciplines of work much before the age of two. From the very beginning they were taught some kind of respect for their master and schooled in simple stable habits, such as getting used to being tethered by a halter, or having their feet picked up for examination and eventual shoeing. They had to learn that man was master not enemy; for if a young horse got the idea that the horseman was a threat to him it might take months of patient work to get him to change his mind. Many of a horse's reactions which appear violent or aggressive are often brought about by fear as a result of an early experience for which they were ill-prepared. Gentle but firm handling of foals was always a good investment. In fact, although horses are always described as 'being broken' to harness, the breaking of a horse's spirit is the last thing you are trying to achieve. It is more a process of gentle persuasion, as stated by the ancient Greeks: 'Horses are taught not by Harshness but by Gentleness.'

By the time of its second birthday, the initial phase of a foal's working life was beginning, and it would now be the horseman's job to get the young colts and mares used to the feel of the harness, the jangle of the chains, the metallic taste of the iron bit in

Shire mare and two foals.

their mouths, the feel of the pull of the collar on their shoulders. Work itself could wait until the horse was fully grown, at about the age of four, for to start heavy work too early was to guarantee a horse that would become crippled before its working years were done.

Most horses succumbed to the ways of work, but not all, and individual horsemen developed reputations for being able to tame a horse that had proved wild and unmanageable in the hands of others. In Victorian times such men had celebrity status and carried an aura of magic about them. But if you look closely at their techniques, which often consisted of tying horses with ropes till they fell, then it becomes clear they were partly demonstrating to the horse that man was in charge

Giving winter fodder to dairy shorthorns in 1943.

and had complete power over him. Once a horse believed that and understood it, the job of schooling was more than half done.

On the farm, yard work continued throughout February. The routines which had been established back in the autumn as livestock came in from the fields to winter in strawed yards, were now well rehearsed. Mucking out, feeding, grinding corn, chopping mangels, hauling sacks of chaff from the granary to the stable, cutting slices of hay from the stack and replenishing the hay racks of horses, cattle and sheep - all these tasks ground on through the month. It was often said to be the month in which winter's back was broken, although the intensity of the manual labour at this time of year might just as easily have broken the men's backs.

If all other jobs failed, the continuing work on the drainage of the land could always find employment for farm hands. The draining of land is the least appreciated aspect of farming, to a layman's eye. He will give no thought to why those deep ditches run round the edges of the fields, or why the men spent many of the long winter months working their way along them, slashing the overgrown hedges, digging out the silt, keeping the water flowing.

But any farmer who knew his business recognized that without proper drainage, particularly on the heavier clay lands, there was no chance of growing a decent crop of anything. A waterlogged soil was of use neither to seed nor farmer - the first would rot and the second be ruined. Too much water in the soil at springtime and the first, warming rays of the sun would be wasted in evaporating the surplus instead of warming the land and bringing the seeds, or the grass, into growth. There were many methods for getting the water away from fields, but in the end it was down to manual labour nearly all involving a variety of spades and shovels. At its simplest, a trench was dug, sloping gently to the ditch, the trench then filled with cut branches and covered to

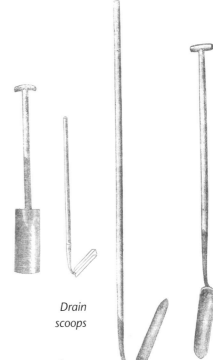

Drain scoops

Clearing out a drainage ditch after cutting back overgrown trees.

form what was called 'bush drainage'. More robust was the laying of clay pipes, but this required more skill, perhaps, and the use of accurate levels to ensure a proper flow of water. Whatever method was employed, it all ended up with a man and a spade spending long, back-breaking hours in winter weather. What a blessing it must have been when mechanization arrived, and the mole plough, dragged through the soil to create a subterranean tunnel, made the ditching spades redundant.

At this time of year the horses will be working at their hardest, readying the land for sowing the seed, which could happen as early as mid-February on lighter soil in warmer districts. The greatest fear of any horseman, where his horses are working at their limit, will be illness. The battery of modern veterinary methods was not available for diagnosis and treatment, and the only cures available were those handed down from horseman to horseman, many relying on as much magic as medicine for their effectiveness. In many parts of the country where flints were to be found in the fields, those which had been worn down by the weather till a hole appeared in the middle of them were said to have special properties, deeply trusted by horsemen. The hag-stones, as they were known, were collected and treasured and simply hung on a beam in the stable directly over where the horse stood. Ask a horseman why, and it's doubtful he would have a clear answer. Often, they claimed it was to prevent the witches riding the horses in the night.

This clearly ridiculous theory is, however, worth a little examination. The horseman's greatest fear is to walk into his stable in the morning and find a horse in an uncontrollable sweat. It can often be a sign of colic - a twisting of that meandering gut for which there was no cure. Horses which were in hard work and were consequently heavily fed were more prone to colic. Of course, if the witches came in the night and took the horses out riding, the horse would show the same sweaty symptoms, which no horseman wished to see. So, if he was honest, a horseman might admit that the hag-stone was there to bring him luck and keep the colic, and not the witches at bay.

If superstition failed them, the old horseman turned to herbal remedies to cure all ills. Not that you could enquire too closely of a horseman precisely *what* he used to treat his horses for he would probably reply in a well-practiced and evasive way that he, 'didn't use anything special at all, but he'd heard of men who did.' Horsemen kept their secrets, careful never to write down the recipes for fear they might be stolen. The recipes, of course, were valuable especially if they'd proved effective in the past, and a little deception was called for if one horseman wanted to find out another's secrets. It was not beyond them to follow each other to the chemist's shop, and ask for 'same as that man's just had, please.'

To take a few examples of the herbs they might have used: common agrimony with its small, yellow flowers could keep horses in condition; should a horse be suffering from cracked heels, then a strong infusion of this herb was said to be a cure. A condition which was well known, and might have shown itself at this time of year, was for a horse that was in heavy, constant work to go off its feed. It became, in effect, too tired to eat. The cure, the old horsemen believed, was the leaf of the elecampane herb. Horehound, likewise, kept horses 'on their food' while belladonna, or deadly nightshade, was mixed up to a thick syrup and spooned onto the back of a horse's tongue to cure it of a cough. Then there were the oils, used as much on the men themselves as the horses, it seems. The recipe for one of these oils included egg yolk, spirits of hartshorn and white vinegar. Not all ills could be cured by herbs, though, and the tale is often told of the farmer who had a stable full or horses which all went down with a glandular complaint in the middle of winter, at just the time the farmer's need was greatest. He gave them no medicine but turned them onto the sparse meadows where they immediately took whatever few blades of grass they could find. Within days they were restored to full health. It's not difficult to see how, in a less scientific age, there could appear to be much magic surrounding the care of cart-horses.

Good protection for ewes and lambs with plenty of feed in the stack on the right.

Mindful of that old saying about having half your corn and half your hay still in reserve at the beginning of February, this is the time of year for the farmer to be making some careful calculations. Optimism is always dangerous in farming, but if his haystacks are looking a little depleted, he may comfort himself with the fact that he's 'due an early spring' But there are never any guarantees and certain members of the farm are getting ever hungrier, in particular the lambs.

For the first weeks of life, the lambs have grown fat on their mothers' milk. To enrich it and guarantee a good flow, the ewes have been fed on hay, turnips and possibly some of that precious linseed cake which the farmer prefers to keep for his in-milk cows. But now the lambs themselves are beginning to take an interest in solid feed and instead of hanging back at the troughs while their mothers feed, they too will be pressing forward to share in the rations.

What's good feed for the ewe is perhaps not best for the lambs and so the shepherd will start to set up his 'creeps.' At its simplest, a creep is a narrow gateway through which a lamb may pass, but its mother can't. On the far side of the creep, accessible only by the

Taking mangels from the clamp for winter feed: 1942, East Yorkshire.

1941: students at the Wallingford Farm Training School are slicing mangels for sheep.

lambs, is their ration. Without the barrier provided by the creep there is no doubt that the selfish ewes would not only have consumed their own ration, but their lambs' as well. Motherly love amongst sheep does not go so far as to allow a ewe to hang back to allow her lamb a good feed. Remembering that the lucrative spring lamb market is rapidly approaching, the shepherd feeds his lambs on rich, clover hay which will have been cut, or chaffed, into shorter, more easily digestible lengths. There might be some oats provided too.

Apart from depleting the precious winter food stocks, the increased work load on the shepherd can start to take its toll. If the sheep are being fed turnips in the fields, he will have set up a system of fencing, perhaps using hurdles, to confine the flock and ensure they 'clean up' before being allowed to move on to a fresh patch of turnips. If

Sheep creep

Sheep feeding on roots. In the background a horse-drawn timber drug is making its way up an incline.

he didn't manage their grazing and failed to control their wanderings, they would idly graze the field taking a turnip top here, another there, and much would go to waste. Unfortunately, the way the turnip grows, and how a sheep's teeth have been designed, don't make a perfect match. A sheep will certainly nibble away at the top of a root, but once it has been gnawed to ground level, it will turn its back and set to work on another, leaving almost half the valuable turnip in the ground. This provides one more job for the shepherd who takes up his turnip picker - a long-handled hoe - and walks the rows, stabbing his picker alongside the buried root and flicking it upwards till it lies on the ground, ready to eat. The hard work having been done for them, the sheep will happily return to finish off the job.

And while he walks amongst the flock, a good shepherd's eye will never stray far from his sheep. These creatures are prone to all manner of illnesses, far more than a horse or a cow. And at this time of year, when the flock is confined and the ground can be wet, any sign of a limping sheep is a signal that foot-rot may be about to spread through the flock. It has been a long winter for the shepherd, and like many on the farm he will not be sorry to see the end of February.

March

What a fickle month on a farm! March can never seem to make its mind up whether it belongs to winter or is attached to spring. Whichever it chooses to be, you can bet it will not be the one the farmer wants. Much, of course, will depend on the soil the farmer is blessed with. The heavy clay lands may still be sodden and unworkable, whereas the lighter, sandier soils can lose their moisture quickly, even at this time of year, and spring can be held back for want of a good shower or two of rain.

A few dry March days, it was said, could make or break the following harvest. Seeds need to be sown in this month if they are to get the full benefit of the growing season. But seeds need dry, prepared and warmed land in which to flourish and achieving all these conditions depends, as ever, on the weather.

If the ploughing and the cross-ploughing are not finished, then there will be an almighty rush on any day the sun shines or a drying breeze blows. The farmer will pace the yard impatiently, wondering why the horses are not in the fields, why the seed drill remains stationary in the shed, why the harrows are not working their way across the fields instead of gathering rust. A farm bustles in March and tempers become short when the weather provides a break for a vital job to be done, and neither horse nor machine are ready for it.

Of all the farm implements, it may well be the harrows that get the most use this

Scottish iron harrow   *Howard's self-lifting wheel harrow*

45

month. The harrow is a crude, unsubtle, yet extremely versatile farm tool. It is essentially a collection of spikes, fixed to a rectangular iron frame, built for dragging through the soil or across the meadows with the sole intention of stirring things up a bit. Built in different sizes and weights, the could be termed 'clod harrows' or perhaps the lighter 'seed harrows', or 'chain harrows' which are more flexible, lighter, and made for hauling through meadows.

The man who gets the job of grass harrowing can count himself lucky. His task is to take his harrows and a pair of horses to the meadow, and stir it about a bit. The process is not unlike combing your hair; the harrows force their way through the matted tufts of grass, break them up, let the air and sun see the roots, and generally waken them to the fact that spring is about to burst. But the harrows do more than that. There will be patches of droppings left at the end of the previous year which

'The man who gets the job of grass harrowing can count himself lucky': Hampshire, 1940.

Preparing a tilth with an iron harrow. The long, sharp tines take a deep hold on the ground and make hard work for the horses.

need scattering. Doubtless, a mole or two will have left their mark and the hills need demolishing. And there's the matted, dead grass which has withered in the winter and now lies like a crust across the ground. The sole object of harrowing is to get the warmth of the sun to the roots of the grass, and at this the harrow can be deadly effective.

From the point of view of the man who walks behind the harrows, it can be an intoxicating experience. For a start, this is dry weather work and so there is the pleasant sensation of a warm breeze on his cheek. But there's also the heady smell of growing grass, broken by the harrows, releasing its scent. It is the true smell of spring. The horses will feel the warmth too, and may put on the first sweat of the season if the horseman doesn't let them stand a moment at the headland 'for a wind'. Taken together, the smell of the grass and the scent of the working horse are enough to lift the spirits of any horseman jaded by a long winter.

Light horses, probably Highland ponies, used on harrows at Rickmansworth, Herts in 1942.

No one who is working in a field with a pair of horses can consider himself to be alone. A good horseman can communicate with his horses as effectively as with his workmates. Equally, a good cart-horse knows the voice of his master and will respond. Horsemen developed a verbal shorthand, which varied across the country, but in East Anglia the same words were used (with local variations) to give the horse its marching orders. Learnt at the vital stage of being broken to harness, the horse knew that if he heard the words 'Cup … cup … cup … ' then he should turn to the left. This is probably an abbreviation of 'come to me' for it was the practice that whatever the job, the horseman always stood on the horse's near side. To turn to the right the horseman called 'wheeesh …' in a flowing way, like a traction engine slowly letting off steam. Reverse gear was found by calling 'Hu … back' while at the same time giving a gentle pull on the cord with equal pressure on both ends of the horse's bit, unless you wanted to be clever and steer him backwards, in which case the horseman

Breaking up clods with a Cambridge roll and harrows.

played one side against the other. But a good horseman would always be talking to the horse, not just when it needed a command. He would be goading it on, noticing if it craftily took an opportunity to slack and then urge him forward. He might offer a few words, no more than a grunt, to reassure the horse that he was still there, and tell him, now and again, that he was doing well and was indeed a 'good ol' hoss'.

There might appear to be little skill in driving a set of harrows, providing the horses are well schooled to the task, but a good farm worker brought some pride to every job and a properly harrowed meadow would be left with broad stripes, as straight and parallel as if made by a modern mower on a

Cast-iron land roller

bowling green. But the display of horsemanship sometimes went further than that, and if there happened to be a tree inconveniently growing in the field, then a proud horseman would jiggle his horses, sending them in circles round it, until it looked at first glance as if he had driven straight through it. After the harrows came the roller to crush the clods and level any remaining molehills. This job was a good investment for although the roller, if driven slowly enough, would indeed help push the grass roots firmly into the soil, the farmer knew that come haymaking time when the clipper would be driven across the field to mow the grass, any bumps and lumps would be transmitted directly through the machine and make for a bruising ride.

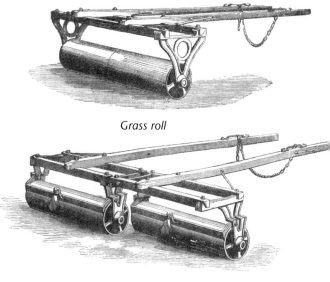

Grass roll

General-purpose roll

The man on the 'clod harrows' had a less attractive task. If the ploughing had been finished early enough in the season to allow the frost to work its magic by freezing and unfreezing the moisture in the soil, the clods would have crumbled to a tilth on their own. But this required everything to go to plan, which it rarely did. Inevitably, some fields were left with hard clods which had to be worked down into fine soil before the seed drill arrived. The heavy clod harrows were the first tool; working the soil one way and then the other, backwards and forwards, horses and men stumbling over lumps of earth which could be the size of boulders. This was when they prayed for the drying winds of March for unless the clods were dry no harrow would split them apart. Too dry, of course, and the harrow couldn't touch them either. Then, the roller was called for and the farm worker told by the farmer in no uncertain terms that unless he drove the roller as slowly as he could, till the horses were nearly creeping along rather than clipping across the field, then he was wasting his time - instead of breaking down the clods, he was merely pushing them into the earth and the field was no more ready to receive the seed than it had been when he started.

The farmer played a waiting game with cloddy fields. A light shower of rain and in would go the roller. Too much rain and the horses would be brought home. But a good downpour followed by a strong, drying breeze and it would be perfect weather for the harrows. One way or another the soil had to be made ready for drilling because next winter's feeding depended on what was sown now.

The wheat, sown at the back end of the previous year, will be coming into strong growth now. But that doesn't mean it escapes the harrows. It will already have suffered at the hands of the sheep who may have been sent across the wheat field as late as November to chomp away at the tops of the young growth. To then send in the harrows to knock it about some more seems almost cruel. The farmer knew, however, that in the case of wheat, cruelty really was kindness because of wheat's remarkable ability to replace a lost shoot with two more. In fact, the more damage done to young wheat, the better the crop, and so the farmer feared neither sheep nor harrows. Both together were as good as money in the bank for a cereal farmer.

No farm could exist in isolation, and no single farm could provide all the skills needed to make a living off the land. Farms were an important part of communities, each providing all the services which kept the main industry going. Few in a community were more important than the blacksmith. His was a two-part job: not only did he use his metalworking skills to fix ploughs and harrows, but kept the horses' feet in trim and provided the shoes they wore. It is a remarkable fact, but a horse without shoes has half the pulling power of a horse fully shod. Not all horses were given expensive shoes, but unless the unshod ones were employed on light work, and never on hard or rough land, then their hooves would become chipped and worn and lameness would follow. The expression 'no foot - no horse' was well appreciated by a farmer who had nowhere to turn for motive power other than to the stable.

The blacksmith's forge was strategically placed, usually near the heart of the village as a gathering place where workers from different farms could exchange disgruntled moans about their employers. In hilly country, especially where the winter weather often brought ice and snow, the forge might be at the foot of important hills so a horse could be brought to the forge in winter to be fitted with special 'frost nails' which would give it grip as it attempted to haul its load up the slippery hill. The blacksmith too was a man of secrets, often as well versed in the dark arts of horse medicine as the horsemen. In some communities, he was effectively the veterinary surgeon and he would be no stranger to the oils and lotions which could cure all manner of equine ills.

On March days, when the weather was less kind, it was no hardship for a farm worker to take a horse to the blacksmith to be re-shod. The heat from the blacksmith's hearth could be guaranteed to warm the cockles of any chilled heart, and through the smoke, the fumes, and the sharp scent of burning hoof, there could be much gossip to exchange.

The village blacksmith.

The methods the blacksmith employed at the anvil appear crude at first glance, but every hammer blow carried with it years of experience. To turn an iron bar into a shoe that is a perfect fit for a horse still calls for immense skill, not only in the understanding of the ways of metal, but in appreciation of how a horse moves, and how a well-made shoe can make its motion easier. There are as many tales told as horseshoes have been made, of horses which were condemned to the knacker for their lameness only to be revitalized by the skilful blacksmith who saw that a specially forged shoe could ease a strain. But ask the blacksmith how he does it and you would rarely get an honest reply. Throughout the countryside, a working man was only as good, or secure in his job, as the secrets he knew.

The chime of the anvil was as much a background to village life as the striking of the church clock: a blacksmith judged the quality of an anvil by the purity of the note it made when struck. Built to stand on a block of elm, complete with its beak and

Inside the smithy at the stables of a large establishment.

throat and hardie hole, it remained a mystery to those who had no appreciation of the arts involved. But it was well understood that the blacksmith could often work magic, and that was the one thing on a farm that was appreciated. So there he stood, strong in back after years of bending and supporting the full weight of the horse's foot. Broad of chest, arms like tree trunks, the blacksmith was a major figure in the farming community in many ways.

If the oats and barley were not sown by the third week of the month, then an air of desperation would set in. The growing season for spring sown cereals was short and any delay led only to a poorer harvest. It's not as though any of the modern options were open to the traditional farmer if a crop should fail: there was no quick phone call to a feed merchant who could supply any shortfall and farms stood or fell by the quality of their harvest, which was why no step along the way, from ploughing through harrowing and to eventual drilling, could be taken lightly.

Before the invention and development of the mechanized seed drill which delivered individual seeds into the soil, allowing them to drop into neat, parallel rows, seed was sown by hand in a way that would have been recognizable in biblical days. Although most seed sowing was done with the aid of some kind of machine by the beginning of the twentieth century, the method of broadcasting still survived. It had two major strengths - it cost next to nothing and couldn't break down. Against it was the inability to control precisely the amount of seed that was sown, and where it went. A special sack, made of linen, was worn around the neck like a sling which allowed a bag to form across the chest. Seed was held in this bag, supported by one arm, while the other arm took handfuls at a time, marching forward, flinging the seed from one side to the other. When it was realized that sowing with both hands was more efficient, the bag was made self-supporting, and 'Stephens' Book of the Farm' of 1886 gives precise instructions for achieving an even spread of seed:

English sowing basket

> *Taking as much seed as he can grasp in his right hand, the sower stretches his arm out and a little back with the clenched fingers looking forward, and the left foot making an advance of a moderate step. When the arm has attained its most backward position, the seed is begun to be cast, with a quick and forcible thrust of the hand forward. At the first instant of the forward motion the fore-finger and thumb are a little relaxed, by which some of the seeds drop upon the furrow-brow and into the open furrow; and while still relaxing the fingers gradually, the back of the hand is turned upwards until the arm becomes stretched before the sower, by which time all the fingers are thrown open. ...*

It continues, but you get the idea. Of course, the seed was not being sown into harrowed soil. The land was still in ridges left behind by the plough. The intention was that the broadcast seed would tumble into the bottom of the furrows, and only then would the harrows set to work. In theory, the seeds would sprout in straight lines, following the line of the furrows.

The seed drill, followed by the harrow.

Drilling in fine spring weather, 1942.

When it came to sowing seed with a drill, the straightness of the work took on a whole new importance. Part of Jethro Tull's innovation was not just to invent a machine which could deliver seed directly into the ground, but also a horse-drawn hoe which could be steered between the rows of growing corn to cut off the weeds. There were no weed-killing chemicals a traditional farmer could employ, even 200 years after Tull, and if he didn't want his crop smothered then the weeds had to be kept under control by hand. Using Tull's horse-drawn hoe, and the developments which followed it, a farmer could run up and down between the rows as many times as he liked, killing the weeds before they took hold and strangled the crop - providing they'd been sown in straight lines in the first place. With broadcast seed, scattered over a wide area, this was impossible.

It's true that perfection in drilling was also a matter of great pride and the driving of the drill was a job for a senior man on a farm. Remember, as soon as the seeds started to sprout, the slightest variation from the straight and narrow would become clear and remain obvious until the crop was so thickly grown that the rows became invisible. God help the man who'd made a poor job of his drilling for he'd be the butt of many a joke while the farm workers stood outside the forge, waiting for the horse to be shod. The jeers would echo the length and breadth of the parish.

April

No other month is packed with life in the way that April seems to be bursting at the seams with the energy of the new spring. The first swallow might return to reassure the farmer that summer will follow, soon to be joined by the martins and the cuckoos. The grass is now truly green, showing none of that pallid imitation of what a rich sward should look like. Instead, it is a deep, vibrant green that reveals it to be full of life. If the farmer applies the old Gloucestershire test, he can be truly certain that spring is here to stay when he can stand on the village green and tread on no less than nine daisies with one foot. Not that he will be wasting his time with such nonsense. He will be remembering the saying that '… work your land in April, and it will work for you'. He will also be aware that there is nothing quite like spring for making a fool of the farmer and all it needs is a shift of wind to the east - not uncommon - and the nights can become cold and the weather turn frustratingly dry just at the time when the grass craves an April shower or two. There can even be snow, of which the shepherd will be only too aware.

It would be a rare farm where everything had gone according to plan, and there will be much work left over from a frustrating March which must be cleared before April can be enjoyed. First, and foremost, is the sowing of any remaining cereals. It's getting late, now, for a good crop of oats or barley. There comes a point in the farming year by which cereals have to be sown or the yield will suffer. Too often, that date arrives too early and compromises have to be made. But only up to a point. If the barley isn't sown by the middle of the month, it might be better not to sow it at all. The hedgerows offer a clue to the timing – 'when the blackthorn blossom's white, sow barley day and night'.

Late sowing of a cereal crop was not always a disaster for a cereal farmer. His oats and barley would most likely be for the feeding of livestock on his own farm, rather than for sale. So if he wasn't expecting to trade his harvest directly for cash, then he

Shearling ewes being auctioned at a Southdown flock dispersal sale in Oxfordshire, 1953.

could probably bear the loss of yield. There were, though, some advantages in patience. If the seed falls into ground which is not only moist but has been warmed by the ever increasing power of the sun, then germination will be all that much faster. Compared with a seed sown early into cold, sodden land, the late crop positively springs into life and, just as importantly, gets ahead of any weeds that are about to sprout. Keeping crops clean of weeds was an eternal battle for the traditional farmer. In fact, the only weapon at his disposal was timeliness, and judging the exact right time to sow crops was one of his greatest talents. Oats sown into warm, cleaned and well-worked soil as late as the end of the month could provide a handsome and clean crop.

The ploughs have almost done their work by now, but on fields where sheep have fed on turnips through the winter, there is one last chance to sow a late cereal, probably barley. The realization that a flock of sheep could be sustained through the winter on a field of turnips was a major breakthrough in agricultural understanding (by 'Turnip' Townsend) and from it sprung several benefits. Firstly, the sheep were not trampling and grazing to within an inch of their life the precious meadows which should be left unstocked throughout the winter in order to make a faster recovery in the spring. Instead, the sheep were folded across the turnip field, confined by moveable hurdles which made their slow progress throughout the feeding season.

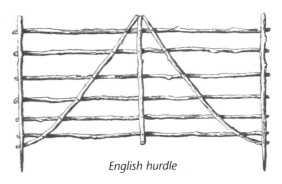

English hurdle

The continuous movement of fences and hurdles was no small part of the shepherd's job. But the sheep, of course, were doing the land a huge kindness as they munched their way forward by leaving a rich covering of droppings as they went rich in vigour-giving nitrogen. The farmer knew that none of the fertilizing effect was to be wasted, so as soon as the last turnip was consumed - which might be as late as the middle of this month - in would go the horses, ploughs, the harrows and rollers, to make a fine seed bed in which the barley could make a speedy germination. Again, timing was everything.

The farmer, gazing at his flourishing meadows with pride, is quite likely to be disappointed at the appearance of the first mole hill around this time of year. Moles were undoubtedly a nuisance, making it harder to mow the grass at hay time, or leading to a difficult harvest if the moles had left their mountainous calling cards in a field of corn. But it was moles in the grassy meadows which caused the greatest consternation because the amount of grass which was covered and killed by a molehill might seem insignificant taken alone, but multiplied a hundred or more times made significant inroads into the quality of the meadow.

The mole is built for life underground and only confronts its enemies - of which the farmer is but one - when it comes to the surface. Owls, buzzards and weasels are natural predators. With its powerful streamlined body, and nostrils that point downwards so as not to clog when digging, it takes a mole no time at all to burrow into even firm soil. Even a mole's fur stands straight off its body so it is not hindered if it has to suddenly reverse along its tunnel. None of this makes catching them any easier. A large estate might well have employed a full-time mole-catcher who used his secret and mysterious ways to entice the creatures into his traps. In case his master should be in any doubt as to how effective he was being, he often skinned his prey and hung the furry little coats in a neat line on a barbed wire fence where the master passed each day. To some, of course, moleskin was a modest crop which earned a precious few shillings for a farm labourer, whose technique was to skin the mole and peg the coat out to dry in the sun. There was also a mischievous use for a dead mole. It was well known amongst wiser horsemen that the smell of a morbid mole had a potent effect on a cart-horse. What the horse feared we can only imagine, but few horses are brave enough to pass through a doorway which carries the deathly scent. It was considered a good trick to rub the dead, furry body up and down the stable door, or even the length of the manger, while the horseman was at work, and then stand and be entertained when he tried to return his horses to the stable at the end of the working day.

April was not without its moments of shame. In the fields over which the seed drills had passed in late February and early March, the corn would now be sprouting and the straightness of the rows was becoming clearly visible. Again, this provided plenty of opportunity for mischief and embarrassment. But more serious would have been bare strips of field, or corners where no seed had been sown. It showed the farmer that his men had not been concentrating and even if the seed drills had been cleaned, greased, and put away for the winter, the farmer might well order them back to the field to finish the job. Even if no such re-sowing was required, the young crop was not left alone: the spiky harrows would be sent in to attack the young weeds, stir the soil, break the crust which might have formed in wet weather, and get some much-needed warmth deep into the soil.

It was also time to plant potatoes, the hungriest of all the farm crops. Since January, the ploughs might have worked the prospective potato field, working the soil to a fine tilth so that the seeds can be planted deep enough. The seed potatoes were planted in baulks, or ridges, often twenty-seven inches apart and drawn with a double-breasted plough which cut a neat V-shaped furrow. None of this was done until the land had received an appreciable dose of farmyard dung - the potato's best friend. No less than twenty-five tons per acre of land might have been applied, all of it carried to the field by horse and cart and laboriously forked on and off by hand. Assuming a cartload might be around a ton, then with twenty-five trips for every acre you can almost feel the ache in the arms of the men as they heaved forkful after forkful of the heavy muck on and off the carts.

There was a school of thought, however, which considered that applying the muck too early in the season was to waste it, and the potato crop was all the better for the seeds being dropped into the stuff while fresh, if possible. So, in some parts, the farmer would work his soil till it was as friable as it could possibly be, draw the land into

Spring cultivations could include the back-breaking work of weeding onions.

baulks and have the muck spread only in the bottom of the furrows. The planters, often women, followed behind the muck cart, dropping the seeds roughly a foot apart. When three rows had been completed, the double-breasted plough made a second pass across the field, this time steering between the furrows, throwing soil to either side as it went, covering the seed and making sure the succulent shoots were unavailable to the birds. For a farmer who took the growing of potatoes seriously, the potato field was run on a strict regime, well managed to ensure a regular supply of dung from the muck cart, ample seed for the planters and enough of them to ensure the horses weren't standing idle on the headland. Even so, the potato remained a gambler's crop for there was never any guarantee of the yield, no matter how careful the farmer might have been with his cultivations, and little certainty of their worth, either.

The arrival of spring brought with it a huge sense of release, especially for the livestock which had over-wintered in strawed yards. Now, the time had come for the cattle and cart-horses to get a taste of spring themselves. The day of turning-out was a finely judged one. On the one hand, it would be damaging to the meadows to be grazed and trampled before they were truly ready, but on the other hand, as long as the animals were in yard, byre or stable, every mouthful of food had to be brought to them, and every

Land girls working potato fields at Bubwith, Yorks, 1943.

feed was one less in the farmer's precious stock. There was, of course, nothing deficient in the diet of an over-wintered animal, but a wise farmer knew there were few things better than a mouthful of fresh grass. Cows will have been the first to have their release, for the quality and the volume of the milk at this time of year was directly related to the quantity of fresh grass a cow could consume. Following them were the fattening bullocks that would still be receiving a supplement of rolled oats, or juicy mangels if any remained in the clamp. Their sole ambition in life now was to get fat.

Then came the horses. Already, they would have sensed what was coming and might have started to dance in their stalls in anticipation. A horseman would have to be careful here for the horses would be as hungry for a fresh bite of grass as a child for an ice-cream on a hot day. A horse would think nothing of pushing aside anything that got in its way. If there were enough men for the job it would be one man to a horse, for to take them out one by one was to risk those who were left behind breaking their head collars mangers in their eagerness for the wide, open spaces. It was best they all went out together.

In all probability, once released a horse would give an almighty buck, flinging its back end into the air and kicking out with both back feet. It was not a gesture of aggression, but of celebration. But to be in the way of those fast-moving back feet was no place for a horseman. So, it was safest to lead a horse to its field and then turn it to face the gate through which it has just come, rather than simply let it go. This way, the horse has to make a turn to escape by which time it is far enough away from you to buck and kick as much as it likes. Not all horses go crazy when confronted with spring grass. Some will approach it in a measured way, but having taken the first, juicy bites will be overcome with the joy of food and freedom that only then do they cavort like foals. Soon, all the horses will be galloping, calling to each other, dashing in circles till breathless and sweating. Only then will they return to their first love, that of fresh grass.

The daily release of the cart-horses didn't make much difference to the horseman's workload. Although his mucking out duties tailed off, horses which had spent the night in a wet field needed energetic grooming to rid them of the mud. The grazing made it somewhat easier, though, for a grazing horse had a bloom to it that a good horseman appreciated, and which made his job a touch easier.

With the spring grass came the fastest flow of cows' milk in the entire year, and that which could not be sold would find its way to the dairy, or the kitchen on smaller farms,

Working horses at summer pasture.

A farm dairy in Wensleydale, 1934.

and the cheese making would begin. For this, every implement would need to be scrupulously clean and the boy, if there was one, would be set to scalding jugs, bowls and utensils, remembering that it is very easy to make cheese but not so easy to make cheese that is palatable.

Traditionally, if making hard cheese, the evening milk would be poured into large, shallow bowls and left overnight, when the cream would rise and be skimmed off the next morning to make butter. From now onwards, fine judgment plays a part. The milk must be brought to blood temperature - certainly not hot - at which point the rennet is added. The purpose of the rennet is to turn the milk sour, which it would do on its own left for long enough, but this way the souring can be controlled. Eventually, the white curds form and fall to the bottom of the bucket, or whatever was being used, leaving the watery whey floating on the top. Straining through a muslin cloth separates the curds from the whey, which usually goes to the pigs - much to their delight.

The soft, floppy curds now form the basis of the cheese. Left as they are, they are a true cottage cheese but it was likely that the farmer's wife would be looking to sell a little of her cheese, or trade for flour or groceries, and a hard cheese would be not only easier

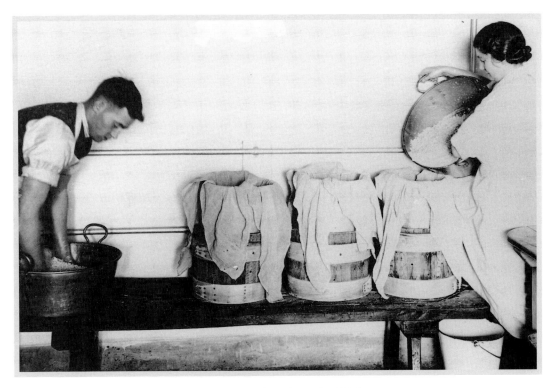

Making Cheshire cheese in 1936.

to store, but to sell as well. So, the curds were pressed to remove the final trace of whey, then turned and pressed again till quite dry. A well made cheese would have a considerable shelf life. Sometimes, they were pickled in brine before being allowed to dry, or simply left in a cool corner of the larder to mature. It is unlikely they would see the light of day till well into the autumn when the ploughing recommenced when for lunch the ploughman would relish a chunk of what was truly a farmhouse cheese.

There was never a time of the year, on the traditional farm, when the farmer could consider himself up with his work. He needed to be planning ahead every day of the year, making the best of what the weather sent him. So it was no good avoiding the sowing of clover if the men and the horses had no other jobs to be getting on with. It might be tradition that clover was sown in May, but the motto 'no time like the present' was one the farmer took to heart.

Barrel churn (butter)

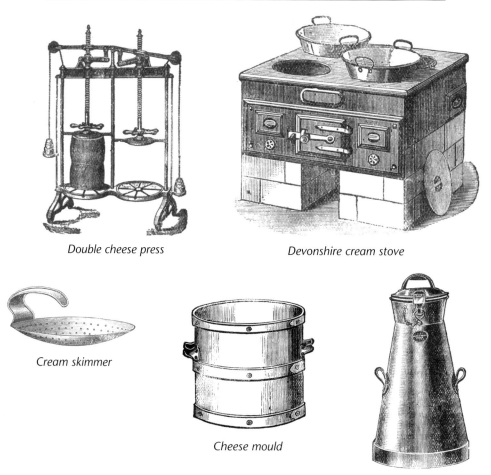

Double cheese press

Devonshire cream stove

Cream skimmer

Cheese mould

Railway milk churn

There were many crops grown as recently as seventy years ago which are never seen on most farms these days, but the traditional farmer knew the value of them. Clover was a versatile crop, for not only did it provide a hugely nutritious feed if grazed or made into hay, but land on which clover had been grown was rich in nitrogen. Growing clover was a natural way of adding fertility to the land and one which no sensible farmer would ignore. Other crops, such as vetches (known as tares in some parts) and sainfoin, also had the same almost magical properties. Of all the crops, these were the last to be sown, for warm soil was essential for a speedy germination, and a severe frost would set them back if they happened to be struck. So, was the farmer to sow his clover in April and risk it, or play safe and wait till May when the weather for drilling might be worse? These were the questions which occupied much of the traditional farmer's thoughts. And while he pondered his course of action, awake in his bed a night, April often brought with it the sound of the first nightingale. Yes, the season had truly turned. He'd sow the clover tomorrow.

May

The vital forces which kept the traditional farm alive have moved underground in the month of May. Here, the growing roots are seeking out moisture and food, and the success of their quest will determine the coming harvest. There's not much a farmer can do now to affect his eventual harvest one way or the other; it is truly in the lap of the Gods. His efforts in ploughing, muck spreading and cultivating will either pay off, or the idleness of his ways in those crucial spring months will come home to roost. If he feels a need to tend his cereal crops, then he can send in the horse-drawn roller to press the shoots firmly into the soil and give them no excuse for not absorbing the food he has provided for him.

Most traditional farmers would have arrived at the month of May with sense of release, but also in a state of exhaustion. The winter was hard work, sure enough, but the efforts of a capricious spring were often enough to bring the hardiest farmers almost to their knees. An old Norfolk proverb sums it up:

> *March will sarch ye*
> *April will try*
> *May will tell ye*
> *Whether ye'll live or die*

The stock yards will be eerily quiet now. The winter bellowing of the hungry bull will not be heard, unless from a distant corner of a well-fenced meadow. The clatter of horses' hooves on the cobbles the only sound as the horses make their way to work, and again at the end of the day when they return. The thick, rich mattress of dung

Chickens free to roam around the farmyard: Bowen Farm, Aldworth, Berks.

and straw, well trampled and rotted now, laid down the previous winter, lies waiting for the men with their pitchforks and tumbrils to cart it to the muck heaps. The only stock remaining might be the loose chickens who have no homes other than in the rafters of the stable, and who are now able to make their way freely around the farmyard, pecking for odd grains. They no longer need to dice with death as the massive feet of the cart-horse fall within inches of where they stand, head buried, scrounging for discarded corn, unaware how close they are to a flattening demise. The cock continues to crow, but there is little there to hear him. It feels as if the circus has left town.

In fact, a whole new show is about to start, and the shepherd will be ringmaster. This is possibly the busiest month of his year. Not only will the winter-born lambs be fattening and seeking food at an increasingly urgent rate, but the ewes who have not asked much since lambing was finished will now be pleading to be sheared. Depending on the size of the flock, there might have been more work than the farmer

Land army girl Helen McEwen shearing with clippers in Perthshire.

or shepherd could handle on his own, and a gang of itinerant shearers would be booked. Watch out! These were boisterous, rowdy gangs of men who travelled from farm to farm; their hot, sweaty and exhausting work fuelled by large quantities of ale. This was, in fact, the very first harvest of the entire year. Not since Christmas has the farm produced a crop the farmer could sell, with the possible exception of some early lambs. It made shearing time a celebration second only to the corn harvest and it was conducted with great mirth and feasting. The farmhouse kitchen would never be quiet as long as the shearing continued.

A more mechanized approach to shearing.

As in every farming operation, the weather is in charge. At some point of the year, usually in May, when the weather warms and the sheep's instinct tells it that it is time to shed its winter coat, the fleece starts to 'rise' from the body. In fact, this is a release of oily lanolin but the effect is to provide a soft, greasy band of wool between the fleece proper and the sheep itself. It is into this dividing line that the shepherd's shears must go. Mechanical shears are a recent invention and even crude, hand-wound ones weren't seen before 1900. Before that the shepherd used scissors, some spring-loaded others merely like a large pair of domestic croppers, but these he could wield with precision and it was the mark of a well shorn sheep that the neat pattern left by the clippers down one side of sheep matched the pattern left on the other.

There was no point trying to shear a sheep while it was wet, so for one night only the yard will have come alive again as the flock is confined, undercover, in case an unexpected shower should fall in the night. The wool, then, was of value (unlike now when shearing makes no profit at all) and so to keep it clean, the farmer would have insisted on the flock being bedded on fresh, clean straw. If he was of the old school he might have gone to the trouble of washing the sheep a couple of days before shearing, but when this enormous amount of effort added little value to the fleece, the practice died out.

The method of shearing a sheep, while keeping the fleece intact and the ewe with all her limbs and teats is almost impossible to describe. The shepherds probably liked it this way, enjoying the secrecy, for it would add even more mystery to their job and make them that bit more indispensable to the farmer. As with the horsemen, shepherds too were only as valuable as what they knew that the farmer didn't. But before so much as a pair of shears can be wielded, the sheep have to be caught, handed to the shearer, and wrestled onto backs where they quickly realize that struggling is a waste of time. The man who does the catching may well have the sweatiest job of the lot. A large farm might employ six or more shearers in order to finish the flock in a day; each of these men might be paid by the fleece and so any interruption in their work was money out of their pockets. If the catcher failed to deliver when the shearer was ready to receive, the air would be blue.

Compared to a shearer using a powered machine, who today might shear a sheep in less than a couple of minutes, a hand-shearer would be considered worth his money if he could clip thirty a day. Imagine the pain in his fingers, the ache in his wrist, after a few hours of that.

It might be the farmer's boy's job to perform the last of the shearing tasks. The precious fleeces must not only be bundled, but cleaned of droppings, daggings, straw and muck. Then they must be rolled in the approved fashion: the fleece, if properly taken, will be whole except for a few clippings which are gathered up and placed in the middle of it. The fleece is then rolled into a bundle and tied with wool taken from the neck of the sheep which has been twisted to form a rope. The rolled fleeces are then wrapped in a canvas sheet to await the wool buyer who will prod and assess them and no doubt offer far less for them than the farmer had hoped.

And as the shearing day comes to a close, the shepherd will be looking to the sky, for a clear, dry day may end with a chilly night and his freshly-shorn ewes will feel the cold. In fact, any of them which are still giving milk to their lambs might be so shocked by the unexpected feel of the weather that their milk dries up in protest. So for a few nights, the farmyard might be in use again before the silence of summer descends on it once more.

Making thatching spars at a competition organised by the Devon wartime agricultural executive committee in 1944.

Not all farm jobs at this time of year commanded centre stage. There was much work going on in the wings - vital work but the jobs on which the spotlight rarely fell. It might be a time to collect ash and hazel to make 'broaches' if a man possessing the skill found himself with time on his hands. A 'broach' (pronounced: *brorrch* in the east of England) is no more than a stick, about three feet long, used to hold the straw thatch to the roof of the stack. Come the winter, when the hay and corn stacks had to be protected from the weather, the efficiency of the thatch covering meant everything to the prosperity of the farm.

The making of the broaches was a skill, not a huge one, but one of those jobs which becomes easy only after a decade or so of doing it. Hazel or ash is cut fresh from a coppice, and chopped into lengths of about three feet. These are too fat, and hazel, being round, might easily be dragged out of the stack by a strong wind. If the sticks are split along their length, not only do you get two for the price of one, but they now present a rougher, gripping surface. To split the stick, the bill, axe or knife is set across the end of the stick and tapped gently so that it enters the wood. It is then levered sideways while at the same time the wood is twisted with the other hand, and the stick splits neatly in two! Like so many farming tasks, what can be described in a few words can take many years to perfect.

Mangels, or mangel-wurzels, can be sown as late as May if the farmer is a gambler, but more likely they will have been planted the previous month and the glossy, pale green leaves will be showing above the ground. The mangel, remember, was a vital crop on the traditional farm for here was a way of providing winter food. The mangel itself is a root crop, resembling a swollen radish which can grow to the size of a football, and swells through the summer to be harvested before the first frosts of autumn. Stored under protective straw, a kind of fermentation takes place inside each root turning the starch to sweet sugars which few animals will not relish. On a diet of mangels, hay and a little cereal, there is nothing on the farm that would not thrive. The drawback with the mangel lies in the way the seed germinates. Instead of each seed producing just one set of leaves, and eventually one root, a single seed can produce four or five. But for a good crop, only one root per seed can be allowed, otherwise you will have an indifferent crop the size of pears.

Therefore one more laborious task in May is to 'single' the mangolds. In other words, men and women took up long-handled hoes and marched up and down the rows, killing off the surplus and leaving just one set of leaves to thrive. It was a job which required no small amount of skill, for the leaves were often tangled and it was

Two Birmingham lads learning the art of hoeing near the YMCA hostel, Stratford-upon-Avon, 1945.

too easy to kill the entire root with a careless flick of the hoe. It was a job to which a farmer's boy might be set, and by the time he was a man not only would he have mastery of it, but he could also perform it at speed. Then, he could join the gangs of hoers, shuffling their way across the acres without an apparent thought in their head, wielding their sharpened hoes with the precision of surgeons.

The hoeing, of course, had a further benefit. The mangel crop, which was far from 'thrown' by this assault, relished the singling and within days grew away with increased vigour. The hoes also got rid of any weeds growing between the plants – one more enemy defeated. But to deal with the weeds that grew between the rows, then the horse-drawn hoe was called for. Seeing this machine at work makes you realize the significance of the invention of the seed drill. For the first time it allowed the farmer to plant his seed precisely where he wanted it, and in the desired quantity. There is, of course, no reason why mangel seed could not be scattered over prepared

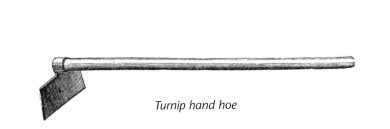

Turnip hand hoe

ground; indeed, the roots would grow, after a fashion, but there would be no true crop because the weeds would soon have the upper hand, and short of pulling them one by one - an impossible task - there was no other way they could be controlled.

But with a horse-drawn hoe it was the easiest thing to run down the parallel rows, killing off the weeds. It called for a steady horse, and was certainly no job for a novice gelding, or a novice horseman. In a young crop, the horse would probably be led (another job for the boy) but when the crop was well up and clearly visible, an experienced horse might be trusted to find its own way between the rows, which they could often do with amazing accuracy. It then became the job of the horseman behind to steer the hoe to compensate for any wanderings and in no time at all several acres of root crop could be cleared of weeds. In fact, this would have been a regular sight throughout the summer months as the hoes went through crops of mangels, kale, cabbages, turnips and potatoes. Never a chance must be lost to 'get on top of things'.

The month might provide the farmer with a chance to take a breath, and simply look around the farm. Fewer jobs were urgent now. Certainly, in the nineteenth century, this was a time to farm with less haste than in many other months of the year.

Horse hoe

But come to the introduction of silage making, and the traditional farming year managed to plant yet another mile post, and an exhausting one at that.

The preservation of grass, by drying in wind and sunshine and then building into a stack for future use, is called haymaking and is the major task of June. But ensilage, where the grass, and other crops, is cut green and preserved, takes place some weeks earlier.

If fresh cut grass was merely gathered together, built into a heap and left, it would rot. Like a giant compost heap it would warm, almost to fire heat, before reducing the grass to a slimy, useless mass. The secret of preventing this happening, and thus preserving the grass and its food value, was to remove oxygen from the process - fresh air was what did the damage.

Silage is made in a silo which could be a pit, a hole in the ground, or a tower. Either way, the exclusion of air from the fermentation process was achieved. Properly conducted, and at exactly the right time of year when the crop was bursting with proteins, silage could be a winter lifesaver for a stock farmer, and sweet relief from a diet of dry hay and cereals for the stock.

The making of silage called for some judgment, though, and anyone who thought it was simply a collecting and dumping operation was in for a shock. Firstly, the horse-drawn mowers would take to the fields, not only cutting grass but crops such as peas, unripe cereals and vetches. Then, the long, hard carting process would begin, forking the cut crop (after a day or two's wilting) into horse-drawn tumbrils which would be led to the silage pits. Here it was tipped and the pit slowly filled. To remove the air, it was weighted with whatever was to hand; it might be covered with earth, bricks and stone, all of which involved yet more carting and hauling. Its advantages over haymaking soon became clear to the farmer. First, and foremost, he was at last independent of the weather. Haymaking needed wind and warmth which were not always guaranteed, but grass cut for silage could take a downpour and still be fit to cart twenty-four hours later. Nutritious crops other than grass could be utilized, such as oats, rye, maize and barley, all cut well before they would ripen if allowed to stand as a cereal crop, and just as good value as feedstuff. A spin-off from this was that the ground where those cereals had been growing could be cleared early, and quickly sown to a winter fodder crop for sheep folding, thereby getting two crops in the year from a single field, rather than one.

Before the nineteenth century, such ways of farming would have themselves appeared intensive, even if benign by modern standards, but the invention and

Almost as tiring as stone-picking: pulling docks by hand, Hampshire, May 1940.

exploitation of ensilage is a good example of the relentless intensification of farming, even if for the very best of motives.

Some traditional practices, however, were of no value whatsoever. Stone-picking is a job that is probably best dead and buried. Land was often allowed to 'rest', or lie fallow, thinking that by growing no crop it would recover its fertility and a better crop would follow the next season. In the meantime, the land was bare throughout the growing season, being continually worked with harrows and cultivators to kill the weeds and 'clean' the land. In the process of cultivation, inevitably stones came to the surface and gangs, usually of women and children, went to fields to pick them up in return for a few pence. If the process achieved anything is doubtful. At the end of the eighteenth century, an experiment was conducted with conclusive results. Under the direction of Arthur Young, the great agricultural writer and observer, two pieces of land were sown with an identical quantity of wheat seed. One piece of land had been stone-picked, the other hadn't. The yield from the stony field was greater than that from the land which had been cleaned.

◆

This did not bring to an end the practice of stone-picking, for the stones which were gathered became important for road and track maintenance. So, to each field one family was usually allocated who arrived with two-gallon pails and a wooden rake with six-inch nails for prongs. As the buckets were filled, they were carried to a heap, twenty of them making a load – according the George Ewart Evans, writing of the Suffolk way of stone-picking. The end of this practice was brought about when cheaper roadstone could be imported from distant counties, and a fruitless task came to an end. The money was poor, the work back-breaking and pointless, for the next time the plough passed yet another fine crop of stones would appear, and again the year after that. The relentless march forward in agricultural understanding left some practices well forgotten.

June

In the month in which midsummer falls, the countryside looks at its best. Grazing cattle and sheep are presented with a green, lush feast in whichever direction they turn, and in the warm sunshine, and the occasional splattering of summer showers, the roots, corn and fodder crops are blooming. The leaves on the mangels are large enough to meet across the rows, excluding the light from falling on the ground between them, and thereby keeping the weeds in check. Likewise, the leafy potato

Potato plants being hoed by hand in June 1941 at Burscough, Lancs.

Earthing up a potato crop in Oxfordshire, 1950.

crop is smothering the ground and making growth of anything other than the potato almost impossible. It was for this reason that the traditional farmer called the potato a 'cleaning crop' for there was nothing quite like it for ridding soil of unwanted pests. When the potatoes are lifted, and the land ploughed, it will be fit for a corn crop, the farmer safe in the knowledge that the land will be as free of weeds as he can possibly make it.

So, if everything seems well on the farm, why is there such an atmosphere of bad temper? Why, as the month progresses, does the farmer spend increasing amounts of time watching the sky and the clouds, pacing the fields, trying to make up his tortured mind about what he could best do next? The answer is simple - it is haymaking time.

The making of hay is arguably the most important time on a traditional livestock farm. The other harvest seems less significant by comparison. If the corn should prove to be a poor crop there will be difficulties, it is true; but at least there is bound to be *something* that can be harvested. But if the weather conspires against the farmer to ruin his hay then it is all lost, and the major part of his winter feed will have disappeared. The hay must be got, the farmer knows, and good hay too. So he paces the fields, judging when to start to mow, praying that he's got it right.

But others on the farm might also be in a tetchy mood. It is, again, a busy time for shepherds. In fact, there was hardly a moment in the farming year when the shepherd could hang up his boots and catch his breath. The sweaty effort of shearing is behind him, true, but ahead lies the dipping, and the daily vigilance which prevents the unpleasant foot-rot, or maggot strike. Sheep are far from hardy creatures and seem prone to attack at almost every time of year. Foot-rot can strike in summer or winter and the shepherd will eye his flock with great care, looking for a limping ewe or one which prefers to graze on her knees, to take the weight off her sore feet. Foot-rot can be contagious so he must act swiftly, pen his sheep, turn up the sick one and with the knife which is always on him quickly trim the rotten part of the cloven hoof till he sees a fresh, uninfected foot.

The maggot is an even worse predator than the bacteria that causes foot-rot, and on wet days, when a sheep's fleece hangs sodden in warm sunshine, there is no better moment for the fly to lay its eggs. Then, aided by the warmth and moisture, the eggs hatch into maggots which use the skin of the sheep as their feed. Sheep are eaten alive this way, which is one of the reasons that sheep are dipped.

It does not do to enquire too much of the chemical mixture in which the farmer immersed his stock for by modern standards they were probably highly toxic, not only to the sheep but to the shepherd as well. Sheep have a sense of impending trouble, it seems, and no sooner will they have been gathered in the pen and sniffed the dip, than they will bleat and cover the floor with involuntary droppings. The pen then becomes as slippery as an ice-rink and it can be a struggle to catch the ewes and persuade them to take a swim through the toxic bath. It is a job that must be done, and a job for which all are thankful when over, and doubtless by the end of the day the men will have been immersed in as much of the dubious dip as the sheep themselves.

If the hay has not been made by mid-summer day, then the farmer becomes increasingly worried. The point of haymaking is to preserve the grass to provide winter feed; but the grass is at its most nutritious at the point just before it comes into flower and then sets seed. At this stage in its life, it is bursting with protein and energy which it intends to use to reproduce itself, and at no other stage is its feed

Maintenance work had to be fitted in whenever possible. This 'Cornish hedge'
was being repaired in June 1952.

value so high. It is one of the curses of nature, however, that the moment when the grass is at its best for haymaking, the weather rarely obliges. It is mistakenly thought that the farmer is looking for hot, dry weather in which to make his hay, but in fact a good, drying wind is just as good as a baking sun. Some would say that the more measured drying effect of a good breeze makes better hay than a scorching sun, and science would support that. The traditional instruction to 'make hay while the sun shines' is only part of the story. Either way, wind or sun, it is settled weather the farmer seeks. A week of fine weather will do, ten days even better, but without a weather forecast to turn to he has only his instincts to trust, and much is riding on them.

Long before the decision is taken, the machinery will have been made ready. On a farm where the horse-drawn mower has been introduced, the task of mowing is made that much easier. Before then, the scythe ruled the hay field and the honing of the blade to the keenest of edges was something over which a farm worker would spend many hours, and in which he would take much pride. There is far more to a scythe than there might first appear. Firstly, the handle - which was known as the snath, sneath or batt - had to be bent to a precise shape. This was usually done by steaming an ash pole to the required shape, or cutting one from the hedge if you were lucky. In some parts a straight handle sufficed. At the bottom of the pole, which was generally the heavier end, was attached the blade, made by the blacksmith in his forge. The adjustment of this blade was no simple matter and an incorrect 'setting' could render a scythe virtually useless. There were many ways of arriving at a correct setting, and in order to function at their most efficient, a scythe should be reset for each person using it. Firstly, the handles, or 'nibs' must be correctly positioned. This is done by setting the heel (the bottom) of the scythe, against the right shoulder. The lower nib is then placed at arm's length. Next, the upper nib is placed the length of the forearm (measured fingers to elbow) from the lower. Now, lay the scythe across the left shoulder, the left hand holding the upper nib, the other nib against the cheek, the heel of the scythe behind the head. Stretch out the right arm, and the position of the blade should be adjusted so that the point of it coincides with the root of the thumb. There were others ways, and many secrets in the setting of a scythe, as in much other farm work. It was doubtless good sport to see a novice attempting to mow

Mowing machine

his first ever field, for there was as much subtlety in the swinging of the scythe as there was in the setting of it, and to swing from the hips and allow the momentum of the blade to do the work was a technique that was hard learnt. And then there was the sharpening. An old hand at scything would be able to discern the precise moment that his scythe had lost its 'edge' and he would draw from his leather belt a round sharpening stone, give the blade a dozen or so precise sweeps with it, and resume his mowing leaving the novice wondering why his work was getting harder and harder as the time passed by.

It must have been a fine sight to see a meadow full of hard-working men, as many as a dozen in a large field, each following the man in charge who was known as 'the lord'. The only sound would be the swish of the blade through

A scythe with the English-pattern s-shaped shaft

Some help from mechanization: a small engine drives the reciprocating blade of the Bamford mower in Cheshire, 1943.

A traditional farming scene: horses and men refreshing themselves in the yard.

tall grass, interspersed with the rasp of the sharpening stone on the blade. On hot days, the work must have been almost unbearable. Indeed, in some districts, they would commence mowing at dawn, cut till the middle of the day, and only resume about five when the cool of the evening was beginning to fall. Then, they might mow till it was dark by which time it was expected that each man would have mown an acre. Scything grass was a 'more haste, less speed' business, and the masters made slow and steady work of it, knowing that in the lengthy race to make hay there was no point in sprinting.

The invention of the horse-drawn clipper made the scythes into antiques, but did not diminish the skill involved in cutting a field of grass. The sharpening of the cutter blade on the mower was just as important to the horses, which were now doing all the work, as it was to the man who had to swing the scythe.

So, the decision has been made, the horses well fed and watered for what will be a heavy, sweaty day's work, and prayers have been said that the settled weather will

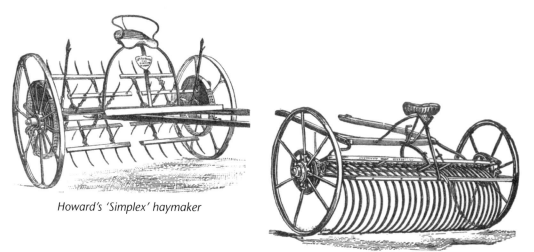

Howard's 'Simplex' haymaker

Howard's self-acting horse-rake

Hay fork

continue. The 'haysel', as it was known in some parts, was about to commence. Rather than drive the horses round the edge of the field, the horseman might well cut a swathe through the grass the width of the mower, about a foot away from the hedgerow. The first cut is generally the most difficult with the horses unused to the feel of the clipper, and with no clear ground to walk on they can tend to wander. The horseman himself will be keeping a keen eye on the cutter, which tends to clog till everything is going smoothly. For all the mechanical advantages of the horse-drawn mower, it doesn't entirely remove the effort from mowing. Jams are frequent as the horses slow momentarily for whatever reason, and the cutter seizes solid with a bundle of grass trapped between the blades. Then, the horseman must dismount, knock the machine out of gear and clear it, while the horses stand ready targets for flies and buzzing insects. Tempers on both sides can get short.

A well-practiced man with a scythe might mow an acre a day; a pair of horses would have no difficulty cutting a dozen if water was brought to them and they were allowed sufficient rests to catch their breath. If the farm was substantial enough to have a stable full of horses, then the team might be changed at the middle of the day and a fresh pair brought into work. This way, the work of mowing need hardly stop. And by the end of the day, the farmer will see his field of grass flattened and lying in rows, or swathes. And now the real work of haymaking can begin.

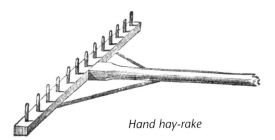

Hand hay-rake

The rewards for the farm worker were poor in relation to the effort required but at least his refreshment was well looked after. The hay field was fuelled with home-made ales and beer, and a man swinging a scythe for an entire day might down several pints. And why not tea? Because until the end of the nineteenth century tea was far more expensive than home-brewed beer, especially in districts where barley was grown and malt could be easily made. Beer became the working man's choice out of financial necessity, and not because of its taste.

For the farmer's boy, June brought with it yet another futile task - bird scaring. The birds were undoubtedly at their most destructive in the spring when juicy, fresh shoots were emerging at the end of the long, hard winter. But in the summer months, despite an obvious abundance of food, the green heads of kale or maize were an overwhelming attraction. The job of the farmer's boy was a simple one: he must walk the rows of growing crops, clapping together two flat pieces of wood to keep the birds flying, and not feeding. This could occupy him for many hours, although with grass lying in the fields, waiting to be carted, along with the entire farm he was on a permanent state of readiness for the first signs of rain. Then, a military-like operation would swing into action with the sole intention of saving the precious crop of hay.

The cut grass might have been let to lie in the swath for a day or two, to allow the

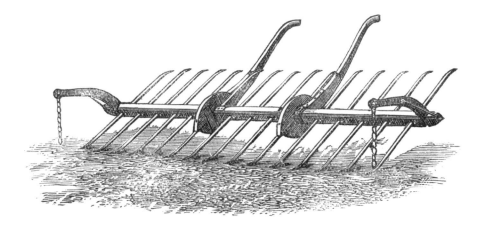

Hay/pea toppler

One of the better jobs for the farmer's boy: giving the horses water.

grass which lies on the top to catch the drying wind. The grass underneath will still be green and as full of moisture as when it was first cut. The secret of good haymaking, the farmer knew, was the slow and steady removal of that moisture so that all the nutrition in the grass became concentrated in the dry hay. So, after a day or so of initial drying, the swaths were turned, the drier grass being turned under, and the still green grass now allowed to face the sun and wind. This, of course, could be a lengthy task if using the two-pronged hayfork. Again, men would have their own forks and be fiercely protective of them. No fork but their own was up to the task, they believed. They got to know the swing and balance of a fork; all part of making a hugely laborious task into an easier one. The arrival of the horse-drawn tedder, or swath turner, relieved the farmer of having to make a long, slow procession down the field, flicking acre after acre of grass by hand. Now, he could ride while the ingenious, multi-fingered device turned the grass for him. In catchy weather, when the rain might strike, the ability to be able to turn grass quickly, and again soon after if necessary, was as good as money in the bank.

If the rain did come it did not mean that everything was lost. In fact, some thought a light shower did no harm whatsoever and would soon burn off, doing the hay no damage at all. But a real cloudburst could do a lot of damage, and if the thunder was heard rumbling down the valley, or black clouds gathered on the distant horizon, then it was time to cock the hay. While still wet, it could not be built into a stack of any size, otherwise it would quickly start to heat up and rot, like compost. But if made into a small heap, say three feet high, then it could safely be allowed to stand for a day or so without the risk of any damage. As soon as the threat of rain was passed, the cocks could be broken open and the drying process allowed to continue. As with most farm jobs, there was more to building the cocks than just throwing the grass into heaps. The grass had to be picked up with the fork and allowed to drape itself over the times. Then, if the fork was flicked over quickly, the grass would fall onto the heap and drape itself over the grass that was already there. This gave the rain no opportunity to seep into the middle of the heap, but instead it would run off the roof leaving the contents dry. It worked. Tales are told by old farm workers of repeatedly cocking and un-cocking whole fields till the hay was 'won', which it invariably was.

In some parts of the country, notably northern areas where rain was more likely, the chances of sufficient dry days to make hay the traditional way, were unlikely. Then, cunning had to be employed, and yet more labour. Even in quite wet weather, good hay was made, especially hay from the leafier crops like sainfoin, lucerne or clover, by building the cut crop around a tripod of poles. Made of straight-growing ash, three poles, eight feet long, were joined at the top with string, and erected as a tripod. Around the three legs were laced horizontal strings or wires. It was on these wires that the grass was laid, forkful by forkful, being careful not to overload the wires or fill up the space in the middle of the tripod, which was destined to become the crucial ventilation shaft. Even in wet weather, the breeze which usually accompanied the rain was enough to quickly get rid of the moisture and in time a good, slowly dried crop of hay could be made.

Haymaking was the main focus of the month, but other parts of the farm couldn't be ignored. The cattle couldn't be told that it was haysel, so they needn't worry about the buzzing warble fly which was determined to attack them. The sheep could not be persuaded that important things were happening on the farm and so they should desist from escaping. The pigs could not be told that feeding time would be that bit later because things were busy. Only the workers on the farm were allowed to have their lives disrupted by haymaking because, in a way, their livelihoods depended on it.

July

The farmer may, if he is lucky and if nature plays the game, manage to take a break some time in the month of July. True, he has yet to finally secure his crop of hay in the stack, but once that is done then there can be a pause before the next major set-piece of the year, which is the corn harvest.

The handle of the pitchfork being used by the boy on the right has been cut down for him.

A holiday for a farmer, though, is not a matter of packing his bags and 'going away.' The idea that a farmer could hand over his land to another while he travelled so far away that it might not be under his critical gaze, hardly caught hold in the twentieth century let alone the days we are remembering. No, a holiday for a traditional farmer, or his workers, might mean no more than a day at an agricultural show, a market visit, a sheep fair. Farming was a complete life and needed no input from other worlds.

But not until every scrap of hay was in the stack could a farmer even contemplate relaxation, and to be torn between an urgent need to gather hay before the rain came, yet leave it to ripen to its best, was a returning dilemma. Every year the farmer was repeatedly faced with these crucial decisions, and every year he fretted that he might not have made the right choice.

The carting of hay was accompanied by much wringing of hands. In scientific terms, the moisture content of the hay was the deciding factor on whether or not it was fit, and in judging this moment the farmer employed the only moisture meter he

Haymaking in Denbighshire, 1945. The horse-drawn rake on the left of the photograph is bringing the crop to the elevator. Four men are working on top of the stack.

A haystack being built at Aston Clinton, Bucks in 1943.

had, which were his own two hands. He walked the fields, kicking the rows of grass, turning them with the tip of his boot. Sometimes he would discover a green, dank patch which told him, literally, to hold his horses and let the grass cook another day or two. But no field was the same all over, and he might take a few steps further and find the grass there dry and sweet as perfect hay should be and now in urgent need of gathering. To finally make up his mind, he gathered some hay in his hands and twisted it till it was the thickness of a rope. He knew instantly if the grass was still damp, or not. If the result was tight and dry with all the grassy scents of summer in it, that told him the final stage of the haysel could commence.

Before the baler, the pitchfork was the sole implement with which the hay was gathered. Large wagons were loaded, one man on top of each, building a stack which would not fall off if the cart hit a hole in the track. On either side of the wagon, the pitchers were flinging hay high into the air, cursing the man on top for not keeping up, the man on top cursing them below for not aiming it at the right spot. And so the hay was gathered in, every last bit of it, ton after ton, moved on the end of a fork.

But this was only the start of its journey, and when it arrived somewhere near to the farmyard, where stacks were usually built for convenience, yet more precise fork-work came into play in order to build a stack which would preserve the precious crop till needed, which might be more than half a year hence.

Stack-building occupied everybody, every horse, on the farm. As many wagons as could be harnessed were employed so that not a minute of perfect weather might be missed. In his nineteenth century manual of farming, Henry Stephens advises:

> *The stack is built by two men who are supplied with armfuls of hay by a number of field workers, whose duty is thus merely not to carry it, but to scatter it across the stack and trample it under foot from one end to the other. The two men each occupies his own end of the stack, and shake and build up the hay before them as high as their breast, from side to side, and from the end of the stack to its centre. After the centre is reached, the women walk and trample the hay backward and forward, holding one another's hands in a row, until it feels firm under foot.*

A stack-builder knew that the secret of a stable structure, built out of something as untrustworthy as dried grass, was to give detailed attention to the corners. Often, a wiser man would shout to a novice 'keep them corner's high, boy!' for if he didn't he would end up with no stack at all, but nothing more than a heap. A haystack was not an organized dump; it was built as carefully as any house with the intention that it would stand just as long if it had to.

The very last load was left on the wagon and this would be used to form the roof. Without a good slope to it, the rain would simply percolate into the stack and ruin the hay. But after building, the haystack settled by as much as several feet and altered shape a little, bulging in places. None of these defects were corrected until some days, if not weeks, after the stack had been built. Only then, when movement had come to a halt, was the final load pitched onto the roof for the stack builder to complete yet another of his annual creations. And when it was done, a finished stack, it was said, should resemble the shape of a loaf of bread: it must be narrow at the base, widening as it rose, the roof a rounded crust overhanging a little to allow the rain to fall away. And if it was anything other than that shape, then no doubt the thatcher would have a word or two to say when he arrived to give it the final, securing touch.

St. Swithin's Day, of course, falls in the month of July, on the 15th. The legend appears quoted in many different ways, but they all carry the same message:

> *St. Swithin's Day, if thou dost rain,*
> *For forty days it will remain;*
> *St. Swithin's Day, if thou be fair,*
> *For forty days 'twill rain nae mair.*

With corn yet to be cut, it is clear how carefully the mid July weather was observed. St. Swithin was a ninth century Bishop of Winchester who asked to be buried outside the church under the 'sweet rain from heaven, and so that the feet of ordinary men could pass over him'. The monks of Winchester had better ideas and on July 15, 871, attempted to move his remains inside the church. But heavy rain made that impossible for forty days and forty nights. Ever since, on the anniversary of the moving of his body, St. Swithin determines the weather for next forty days.

This was far from being the only weather wisdom relied upon by those farmers whose only method of forecasting was to employ their own instincts, or use the lore handed down from their fathers and grandfathers:

> *If it raineth at tide's flow*
> *You may safely go and mow,*
> *But if it raineth on the ebb,*
> *Then, if you like, go off to bed;*
> *With dew before midnight,*
> *The next day will sure be bright.*

> *When ant hills are unusually high in July,*
> *the coming winter will be hard and long.*

When the Moon is at the full,
Mushrooms you may freely pull;
But when the moon is on the wane,
Wait before you pluck again.

Such rhymes may seem nonsense by our modern scientific standards, but who can criticize the old farmer who sees only the waxing and waning of the moon, the rising and setting of the sun, and thinks that all natural things around him are not in some way connected. Indeed, the sowing of crops by the phases of the moon was a practice regularly observed. The waning moon caused nothing to prosper, it was thought, but seed sown on a waxing moon would soon burst into life. There are some who still hold to this belief.

Not all fields of grass, of course, were mown for hay. Water meadows which might be too undulating, damp or crossed with ditches would prove impossible to mow. But this did not mean they could be left to their own devices, for a farmer knew that good, nutritious grass didn't grow by accident. By this time of year, the grass may be failing. It will have been well grazed, trampled, and covered in droppings which, in the case of horses, are often left in the same place making the ground stale and providing perfect conditions in which rank, tough, unpalatable grasses can grow. The ever present thistle is ready and poised to invade any meadow it can find, and July was the best month in which to attack them. Time spent on meadows now paid off throughout the rest of the year. Scythes, or a horse-drawn clipper provided the best tonic for a tired meadow by removing the unwanted top growth, leaving space for fresh grass to grow. If that couldn't be managed, then at least the harrows would spread the droppings and topple the inevitable mole hill thrown up by the cunning little moles who have evaded the catcher's traps. It was the time of year, too, for animals to find themselves in deeper water than they imagined. Proving the old saying that 'the other man's grass is always greener', both cattle and sheep will spend many an hour pressing against fences and hedges to gain just an extra mouthful from a forbidden meadow. Sometimes they go too far and find themselves in ditches, up to their necks in water, eyes bulging as if in a state of shock. Swift action is called for, especially in the case of sheep that are only too willing to give up on the effort of living, and cattle are too precious to be allowed to die. Even when at pasture, the livestock can never be far from a farmer's mind.

In the dairy, warm summer weather brings with it fresh problems. Milk turns sour quickly, and milking calls for scrupulously clean buckets and utensils if bacteria aren't to get hold and quickly turn fresh milk. The farmer's wife, and especially the boy, can cut no corners at this time of year.

The horses probably carried fewer burdens at this time of year than at any other. When the hay has been carted, there is opportunity for them to rest too, with the exception perhaps of a little hoeing of the root crops which would hardly raise a sweat on a fit horse. Of course, a crop of clover hay could leave a field fit for ploughing. Unaware of the science of the nitrogen-fixing properties of plants such as clover, the traditional farmer's knowledge, gained from handed-down experience, told him that nearly all crops thrived if sown on a ploughed-up field of clover. If he had a sizeable flock of sheep, or herd of fattening bullocks, he might decide to take out a little winter insurance and sow a field of fast-growing roots, such as turnips. There was

*A load for two horses. The horseman is obscured by the trace (front) horse
while his bicycle is on top of the sacks.*

Journeying to or from work on a summer's day.

nowhere better to sow than in the ploughed-up remains of a clover field from which the hay had been taken.

So the ploughs might return, briefly, to the fields. This is not really their season, and both men and horses will find it sweaty work. The plough will not have that shine on its breast that comes from repeated winter use; the wheels will squeal where the rust has got at them. Ploughmen may quench their rasp with a little dab of cheese from their lunch bag.

It is time, in July, to buy and sell sheep; both ewes for breeding and lambs for fattening. The sheep fairs, particularly on the downland parts of Britain where the best sheep farming was conducted, were highlights of the traditional farming year, to be looked forward to, enjoyed, and remembered in the following winter months. They were part celebration, with games, eating and drinking. But at their heart was the sound commercial imperative.

Summer cultivation.

A proud shepherd would prepare his sheep thoroughly, for not only was he at the mercy of the auctioneer's hammer but also under the gaze of shepherds from miles around. A working man's pride was often all he had to call his own. The work of preparing the sheep might start as long as a fortnight before a fair, trimming a fleece to give a sheep a better look, washing to improve the colour of the wool, separating the flock into age groups. Shepherds grew more familiar with the mouths of their sheep than with any other part of their anatomy because this was the only way of being certain of a sheep's age. The shepherd knew his own sheep, of course, for he had lived with them for most of their lives, having helped them into the world. But for others, seeking to buy them, the arrangement of a sheep's mouth gave him a cast-iron guarantee of what he might be getting.

This year's lambs will still have their milk teeth, which are easy to recognize. A full grown sheep will have eight permanent teeth in its lower jaw, but begins with only two and acquires another pair each year for three years, after which the teeth drop out. Without teeth, a sheep quickly loses value. Although it may still breed perfectly well, without its teeth it can hardly graze with any efficiency and may not be able to gather enough grass to stimulate a sufficient flow of milk to feed a lamb. A ewe of this age soon becomes a worthless proposition, other than for mutton. So, part of the preparation for market would be a careful look at the mouths of those members of the flock destined for the fair. To make identification easier the sheep were often marked, the system of marking varying widely from county to county, even district to district. But one method was to mark a two-tooth with a dot of dye on the near shoulder; a four-toother has a mark on the middle of its back, a six-toother would get a mark on its near rump, a full mouthed sheep would be marked on its offside rump, and old, broken mouthed ewes, as the no-toothers were known, would get no mark at all.

The sheep fairs were held in the market towns which were at the heart of the grazing parts of the country. On an open field, pens would be erected, built of wooden hurdles, and the air would be full of bleating from the confused sheep and chatter of the shepherds. Describing a downland sheep fair, Richard Jefferies wrote in the nineteenth century:

> *There are thousands of sheep, all standing with their heads uphill. At the corner of each pen the shepherd plants his crook upright; some of them have long brown handles, and these are of hazel with the bark on; others are ash, and one of willow. At the corners, too, just outside, the dogs are chained, and in addition there is a whole row of dogs fastened to the tent pegs. ... One old shepherd, an ancient of the ancients, grey and bent, has spent so many years among his sheep that he has lost all notice and observation; there is no speculation in his eye for anything but his sheep. In his blue smock-frock, with his brown umbrella, which he has no thought or time to open, he stands listening, all intent, to the conversation of the gentlemen who are examining his pens. He leads a young, restless collie by a chain; the links are polished to a silvery brightness by continual motion; the collie cannot keep still - now he runs one side, now the other, bumping the old man, who is unconscious of everything but his sheep.*

It might be difficult for an outsider to gather any clue about what was being discussed. Shepherds often spoke in localized tongues, and had their own ways of counting their sheep. There might be almost as many different versions as there are villages in England, but the shepherds of the Westmorland Fells counted this way:

1 to 5   *Yan, tyan, tethera, methera, pimp*
6 to 10   *Sethera, lethera, hovera, dovera, dick*
11 to 15   *Yan a dick, tyan a dick, tethera dick, methera dick, bumfit*

On it went, in one form or another, till the flock was counted, the numbers chanted at such speed that they were only discernible to someone in the know.

Against this background of shepherd's babble might be heard the chiming of sheep bells, unless removed specially for the fair. On the uplands and fells, but more often on the gentler downs, where a flock might roam over many square miles, often far from the shepherd's gaze, the sheep bell betrayed their whereabouts. But a sheep bell gave the shepherd more clues than simply location. It was possible, with practice, to read the state of mind of his flock. Although he might not be able to see them, if he hears a gentle clunking, unhurried, chime interspersed with periods of silence, then it was likely the flock was grazing gently. If the chimes had more life to them, were more regular, but still relaxed, it was a clue that the flock were on the move, walking in orderly fashion to the next grazing. A cacophony, of course, meant they were being chased by a dog or fox and action was called for.

A cunning shepherd could learn even more about the state of his flock. Bells inevitably had different tones, and once these became recognizable through familiarity they were as useful to a shepherd trying to work out the meanderings of his flock as radar might be to the captain of a modern ship in confused waters. Of a Sussex shepherd, it was once noted:

> *he tells me that when driving his sheep he notes the position into which the various members of the flock inevitably fall. The same impetuous ones are always to be found far ahead of the rest, on the alert for dainty herbage or any possible mischief. The largest of the iron bells he puts on the noted go-aheads. Some say it is impossible to live near a farm where bells are used; but when they are taken off at shearing time, the shepherd 'misses the music' and wonders who could object to it.*

The drover with his sheep.

Back on the farm, a new character is joining the scene - the thatcher. A small farm would thatch its own stacks but a sizeable one would employ the services of a journeying man whose summer and autumn was spent covering hay and corn stacks with impermeable thatch. His needs will now have to be tended to, and also those of the machinery which has stood idle for the best part of a year - the corn harvest gear.

August

For the second time in the farming year, the scythes are swinging. Or at least their smaller cousins, the sickles, might be. Sickles carried smaller blades and had no long handle. Across the fields, the men marched to the gentle swish of the blade through the crop, interrupted by the sound of the occasional grating of the sharpening stone against a blade. There was a pause now and again for a draft of water, beer if lucky, or to curse the heat. It was, at last, harvest time.

It is difficult to imagine how the mechanical revolution must have felt to men who knew no other way to harvest their corn than by cutting it themselves with a blade. Thomas Hennell described the way in which a field of corn was cut:

Wheat ears

The reaper stoops to his work or kneels on one knee, and leaning forward grasps in his left hand the straw near the ground, pushes the blade of the sickle round it and draws it towards him, pushing his left hand over it at the same time, to avoid the cut. After each cut he raises his left hand to clear the ears from those of the next handful to be reaped; and when he can hold no more he lays out the bunch to his side, lifting it high over the standing corn with the ears supported in the curve of his sickle. As he works across the field he clears a strip about six feet wide (or perhaps less if the crop is heavy), reaping across the end of the strip from the outside to the inside and laying out his handfuls together in sheaves, ready for the binder.

Hook or sickle

The 'binder', of course was of the human variety, usually a woman whose job was to secure the sheaf into a bundle. Imagine, then, the relief that must have been felt when the muscle-aching business of cutting the corn by hand was replaced by machine. First came the horse-drawn clipper, which was no more than a mower, and still required the corn to be gathered into sheaves by hand. This was followed by the mechanical reaper which cut the corn and left it, at least, in straight rows. And finally, the culmination of a remarkable period of inventiveness was the reaper binder.

This revolutionary machine was a mower and bundler all in one devilish contraption. It was hauled through the standing corn by three horses, cutting as it went. A 'windmill' arrangement, known as the sail, knocked the cut corn onto a moving canvas which quickly moved it away from the cutter to allow room for more corn. The cut corn travelled upwards until it arrived at the 'buncher' which gathered

The Samuelson reaper, in Suffolk, 1942, delivered sheaves that still required tying by hand.

Using scythes to open a passage for the binder: Ashford, Kent 1942.

it together in a neat, tight bundle. Only when the bunch was big enough did the ingenious knotter work its magic by throwing a string around the sheaf, tying a knot in it, and finally cutting the string before ejecting the corn, now in the familiar shape of a sheaf, onto the ground. Quite remarkable. It must have seemed like a miracle. Or perhaps it was more a curse, for this was the true beginning of the departure of manual labour from the land. But who could blame a farmer for buying one and leaving the sickles to rust in the barn?

Hornsby's string self-binding harvester

The reaper-binder delivered bound sheaves: Nantwich, about 1934.

A crop laid flat by rain and wind in 1948. In the background the tractor driver is struggling with the binder; in the foreground two men are saving the crop with their scythes.

With the arrival of the mechanical knotter, disappeared the true sheaf knot. When the corn was cut by scythe and left on the ground loose, the bunching was done by hand and so was the knotting. To provide a band, an impromptu rope of twisted corn was used. Tying a knot in straw is not an easy matter and the method varied to the extent that in a given field of workers, no two might tie the knot in the same way. Needless to say, each thought their method was the correct one. Some gave it a mere twist to secure it, others tucking one end under the other, some forming a loop through which the other end was passed. It was long, hard work to gather a harvest in this way; a man and sickle might only manage half an acre a day.

When all the corn was finally cut, there was much rejoicing. In some parts, the last bundle to be cut was called 'the neck.' Tradition, probably pagan, said that it must be cut with the man standing between two sickles laid edge to edge, crying as he made the final cut, and shouting, 'I have the neck!'

But in the late nineteenth century, as the mechanical binder marched across the farming landscape, the old rituals were eclipsed by more modern ones and instead of paying homage to the gods of the harvest, the farmer might be more tempted to spend valuable time oiling and greasing his binder, paying homage to whichever god it was that blessed him with this labour-saving machine. He may not have completely understood its workings, but he knew it demanded respect, and oil, if it was to travel the corn fields without a hiccup.

What was easier labour for the men, of course, was harder work for the horses. In fact, the binder became known as the 'horse killer' for the draft - the amount of pull needed to keep it moving – varied drastically from one moment to the next. It was said that at the moment of knotting and expelling the sheaf, the machine required twice as much pull as when it was just cutting. As the knotting might happen every five seconds or so, it was a persistent and tiring jarring on the horses' collars. Most machines, except the smallest, were pulled by three horses, which were changed at regular intervals. A fresh team, if there was one, was always waiting in the stable. Like the cutting of the hay, no opportunity could be missed to cut the corn.

If the acreage of land to be harvested was sufficient to need extra labour, men were

Reaping scythe

employed from outside the farm. Gangs were always available for hire. Their job was known as 'taking the harvest' and the contract between labourer and farmer was written down and strictly adhered to. A typical one, written by George Rope of Blaxhall, in Suffolk, is typical:

We the undermentioned agree to cut and secure all the corn grown on the farm in a workmanlike manner to my satisfaction; make bottoms of stacks; cover up when required; hoe the turnips twice and turn or lift the barley once; turn the pease once - each made to find a gaveller.* Should any man lose any time through sickness he is to throw back 2s a day to the company and receive account at harvest. Should any man lose any time through drunkenness he is to forfeit 5s to the company

Allowances to each man:
1 coomb of wheat at 20s
3 bushels of malt - gift
1lb of mutton to each man instead of dinner
2½lb mutton at 4d a lb every Friday

Barley ears

As the cereal growing season was coming to an end, for the shepherd the sheep breeding season was about to get under way. Rams need to be fit if they are to perform, and the shepherd will bring them in from a far field, feed them some cereal or whatever else is to hand and build their strength and weight for the coming breeding season. He will have calculated when he wants his lambs to be born, remembering the rough guide that for early April lambs the rams start work in mid-November. On hill farms, where winter might hang on till April, late lambs would be what the shepherd wanted; but a low-lying farm, in a mild part of the country, might risk even a January lambing especially if the ewes could be sheltered in a yard as lambing approached.

The ewes too need a little extra feed, known as flushing, but not too much for fat ewes do not easily carry their lambs. There might be a temptation for the shepherd to eye the fields of clover, newly sprouted after the final cut of hay or silage of the season, and allow his ewes a few hours on these. If he does, he will watch them like a hawk. Clover must taste to a sheep (or cow for that matter) like a sweet breakfast cereal. They relish it, and if not prevented will gorge on it like children who don't know

* A gaveller was a labourer who gathered the crop into sheaves or bundles.

when they've had enough of a good thing. The danger is that they 'bloat'. In the process of digestion, the rich clover releases a large amount of gas and if too much is taken at one time, the gas fails to escape. The sheep blow up like balloons and the only remedy is the rapid insertion of a knife into the rumen, at which point they will deflate as quickly as a punctured football. As there is room for error in the stabbing, and the risk of infection following, although every shepherd worthy of the name would know how to do this, he preferred to keep his sheep on clover only as long as was good for them, before returning them to more sparse pasture. The result is that at this time of year, the shepherd and his flock can tot up a considerable number of miles in moving from pasture to stubbles and back again, to and fro, feast to famine.

Pigs, being easier to keep than any other creature on the farm, are not as susceptible to bloat, and it is perfectly true that there is nothing quite so happy as 'a pig in clover'. There is no one better than the pig for clearing his plate, and if the pigs are moved around the farm, feasting a little on clover, gleaning the remnants of turnips and at the same time leaving their copious dung behind them, then the passage of the pigs will have been of great benefit to the farmer. But the cattle want their share of the rich feed too, especially those beasts which the farmer has an eye to sell before the winter sets in. We are approaching the time of the autumn fairs when farmers dispose of the surplus livestock they have no wish to feed throughout the winter. The fattest stock will fetch the best prices and now is the time to let them have the best of the fodder.

The moment at which the corn harvest commenced was less precise than that at which the hay was made. Whereas the farmer judged his hay by wringing the grass in his hands, with the corn he had no precise way of measuring the

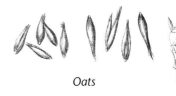

Oats

amount of moisture in the grains, which was the true measure of ripeness. Certainly the colour of the crop played a part, but often as not he would pinch a grain from the head, crunch it between his teeth, and feel the give. If it was soft, almost juicy, his corn was well on the way to being ripe. But if it was as hard as a bullet, it was possibly over ripe and the danger was that the precious ears would be knocked out of the head and lost by the very process of harvesting.

The decision made, the binders were sent into the fields. Like a ghost from a previous age, they were often led by a man wielding a scythe whose job it was to cut the corners of the field to allow the temperamental binder the flow without tangling

Stooking at Bushey Park, Hampton Court, 1942.

in the waving corn. Then prayers were said for the correct functioning of the machine - which had a reputation for having the devil in it - and by the end of the day a decent field of corn would have been cut.

The machine, of course, left the sheaves lying on the field. The binder did not entirely replace manual labour. Now, those sheaves had to be gathered together to form 'stooks', a collection of four or more sheaves in a single structure. Again, what seemed to be a simple task had its secrets and its methods. The object of 'stooking' (or shocking) was to allow the final ripening of the corn, which would have been cut some time before full ripeness to preserve the crop. The method was for the harvester to pick up the sheaves by the heads, one under each arm, bringing the heads together at the same time as planting the base of the sheaves firmly on the ground. Done properly, they will now stand of their own accord. Further pairs of sheaves are now added up to a maximum of ten, all placed in such a way that a clear 'tunnel' runs the

Harvest was a time for all the family to lend a hand.

length of the stook so that drying winds can pass freely through them. Exactly how many sheaves made a 'stook' differed from district to district. Thomas Hennell reports:

> a 'shock' was generally understood to contain six sheaves and a stook, stowk or hattock, twelve sheaves; in the northern counties, twenty-four sheaves, that is, four shocks or two stooks made up a threave. In Cumberland, however, ten sheaves were reckoned to a hattock, or twelve to a stook.

It was, apparently, different again in Cornwall.

It may not seem to matter much how many sheaves there were to a stook, but in the times when the local parson might well expect a share of the harvest as rent, then it was easy to measure the size of the harvest by counting the stooks and calculating

August 1936, Shrivenham, Oxon: two wagons are being loaded with sheaves.

the 'harvest tax' that way.

The stooks were always built in geometrically straight lines; partly for reasons of pride, sometimes to make it easier to drive the wagons between the rows when it was time to cart the harvest home. But on some farms, where the farmer might be keen to extract every last thing he could from his land before the winter closed in again, he might bring his horses and ploughs to the field and draw a few furrows between the rows, turning just enough land to provide a seed bed. Here he would sow a quick crop of fast-growing turnips for sheep to munch in November and December when the pastures were bare and before the field was ploughed proper. Always looking ahead, always peering round the next seasonal corner was the traditional farmer.

How long the shooks would stand in the field varied from crop to crop. Wheat might not stand long; oats on the other hand traditionally 'heard the church bells three times'. In other words, a good fortnight was needed for the oats to ripen and the green straw to lose its moisture. Time waiting to bring the harvest home was not wasted: stack bottoms needed building. As in the stacking of hay, the corn needed careful preservation too, and the base upon which the stack was built had to be carefully considered.

The corn had two considerable enemies. The first was the weather, and rain had to be kept off the crop while at the same time air was still permitted to flow through it. But vermin had to be kept out too, for what better winter shelter was there for a hungry rat than in the middle of a warm stack with as much food laid on as even the hungriest rodent could wish for?

In some parts, staddle stones were used to raise the base of the stack off the ground, but in most parts they took their chances and built the stacks on the ground. Trimmings from the previous winter's hedging work might have been saved, and together with large amounts of straw, a large base would be laid in the precise place the farmer wanted his stack. Sometimes it might be in the field where the crop was grown, but much depended on where the thrashing machine could be set up, and a wet field in winter was no place to be hauling one of those multi-ton monsters. Often as not, the harvest was brought home to stand near the farmyard.

Sheaves were pitched high in the air, falling onto wagons where they were built carefully into loads which wouldn't spill at the first pothole. Back at the farmyard, the stack builder, who was probably the man who had built the haystack a couple of

Loading sheaves in Sussex, 1942.

In the harvest field.

months before, was to be found in no better temper. He cursed the pitchers for throwing the sheaves one way, and cursed them harder if they threw them another. His building of the stack was instinctive, and the humiliation of building one which did not survive the rigours of the winter weather was never far from his mind. He built the corners high, as with hay, and judged when to start on the roof, and how steep the pitch should be, leaving the last few on the wagon to top it up when the stack had settled.

In the fields there was a sense of a great annual battle coming to an end. The very last load was known as the 'horkey load' and was crowned with a green bough, sometimes topped with the best and largest sheaf of all to be set on the very top of the stack with great cheering.

But the real celebration was yet to begin. The farm workers wanted everyone to know that harvest was safely gathered in, and went from house to house, hammering on every door, 'hollering largess' and demanding beer or cider, which was freely given often with meat or bacon. There was plenty to celebrate on the night they could truthfully claim that the harvest had been won.

September

For the first time for many a month, the farmer's boy leaves his bed and notices that the sun is not yet up. Below, in the kitchen, a fire may now be needed to ward off the first damp signs of the coming autumn. There is a distinct change; nights are colder, some days warm and dry while others can be submerged in downpours. There is no knowing September. The old saying that this month, 'dries up wells or breaks down bridges' speaks of the fickle nature of this time of year.

With the harvest mostly gathered with the possible exception of some late spring barley which has yet to face the binder - there is a sense that the growing season is grinding slowly to a halt. True, there will be fresh greenness on the meadows from which the hay has been taken as new growth bursts into life, fortified by the autumn rain; and the root crops still have plenty of growing to do before they are lifted in October and November. But the trees and the hedgerows are looking tired, fit to drop their leaves. And if the animals could speak, they would tell you that for all its vivid colour, that new grass is a pale imitation of the fortifying grass of spring.

The carting of the last sheaf from the corn fields was a time of celebration, as we have seen. But it was also a signal, in poorer times, for the most poverty-stricken of the community to begin their labours. It was the custom that once the last sheaf of corn had been removed from a field - and not a moment before - then the field became available to all who wished to glean it for the morsels of corn the harvesters had missed. What was known as a 'guard sheaf', a single sheaf of corn, was sometimes left by the field gate and not until it was removed was the field available for gleaning. This was tedious, back-breaking work, scouring the stubble for ears and grains of corn. But when agricultural wages were pitiful and families were large, then the corn that might be gleaned at the end of the season could be sufficient to keep a family in bread

Women gleaning near Bishops Stortford, Herts, September 1946.

for a winter, without which it might be one more step on the road to starvation.

Not that farmers made any efforts to leave much behind them as they harvested, despite knowing how vital the leftovers might be to others. Not only was every last sheaf carted but the fields were drag-raked with horses, effectively gone over with a fine-tooth comb, to gather up the very last fruits of harvest. The gleaners could have the rest, if there was any and if they could find it.

The sounding of the church bells was the signal for the gleaning to begin, usually rung at eight in the morning to signal that the fields were open, and again at six in the evening to mark the end of the gleaning day. This service had to be paid for by the gleaners, and the bell ringer's name was posted by the church door so that each family may pay him a few pence for his services.

The coming of the binder and the abandoning of the sickle made life ever more difficult for the gleaners. Before, it was easier for the harvesters to miss a few ears of corn between cutting and hand-tying. But the ruthless binder seemed to get nearly every one, and the gleaners must have prayed for a patch or two to have been left where the machine jammed, or the horses jibbed, leaving a few stalks uncut. Gleaning could be an uncertain way of feeding yourself through the winter.

Nevertheless, a family expected to gather at least a good handful or two of corn by the end of every day, and it was carted home either in their hands or stuffed alongside the baby in the pram, Once home, it was threshed using a hand-driven machine, if such a thing existed in the village. Otherwise, they hit it repeatedly with a flail till the ears of corn submitted and dropped away from the straw. Carefully placed in a sack, the precious corn was taken to the local mill where the miller scrupulously kept each family's flour in separate bags, avoiding argument. With so much at stake, dissent was common at gleaning time. Disputes arose when parish boundaries crossed the middles of fields and over-zealous families crossed the border into another's territory. Soon, though, the rains came and brought the gleaning to its natural end as the remaining sodden ears of corn spoiled. It was now that the family looked at its paltry bag of flour and wondered how they might make it last.

For the horseman, this was the time to wean foals born earlier in the year. It could cause much commotion. For much of the summer, a foal may have worked alongside its mother, trotting beside her as she did her work, hardly leaving her side, taking milk from her whenever it needed. But as it adopted an adult diet, and as the time approached to be thinking about getting the mare in foal once again, there was no putting off the weaning. A gradual process of removing foal from mother might have started before harvest. Instead of running alongside her mother all day, the foal might have been removed for the working hours, and reunited at the end of the day. At first, the mother calls pitifully for her offspring, and the foal calls back. It requires a dominant horseman to prevent a mare from running back to the stable on hearing such a call. Then comes the day when the foal is removed entirely; if the mother has any milk left it is removed by hand. Within a week the mare is dry and the foal seems to have forgotten they were ever related.

The haystacks will now be well settled and if they have maintained their upright stance, then the farmer has nothing to fear from the coming winter. Likewise the corn stacks will have compressed and assumed their resting shape. It is to be hoped the stack ends up with a decent look to it, for there was no rural entertainment like criticizing another's work, and a lop-sided stack was yet another excuse for merciless leg-pulling either in the pub or at the blacksmith's forge. It is also one more

Covered corn stack

opportunity for the farmer to moan about his labourers. H. Rider Haggard wrote in his diary, at the end of the nineteenth century:

> *In walking through the yard I noticed that on my farm at any rate - and others complain of the same thing - the stacking is now very inferior to what it used to be. Nearly every stack leans this way or that, and is propped up with boughs of trees and pieces of timber; also they are roughly and untidily built. The old skill, upon which he used to pride himself, seems to be deserting the agricultural labourer. This is a fruit of the bad times.*

It was always the farmer's role to complain, and his man's to be on the receiving end of it. It was a rare farmer who was entirely pleased.

If all the stacks were not thatched by September, then the farmer ran huge risks. It is the month of equinoctial gales which bring rain and wind, both of which can do great harm to an otherwise good crop. It was a good job for damp mornings, for the thatching straw took up the moisture in the air and became less brittle when worked and twisted.

Stack ventilation

The straw itself (usually wheat straw, for oat straw had some feed value and was sometimes used as a substitute for hay in feeding cattle in winter) was of considerable length. If thrashing had not yet taken place, then some straw would have been held back from the previous winter. Unlike the straw which emerges as waste from the rear of a combine harvester, thrashed straw is long and largely whole. What the modern farmer considers waste, the traditional farmer prized hugely.

Firstly, the straw was drawn from the stack. A broad pair of thatcher's hands grabbed a sizeable chunk and swiftly pulled as much as they could hold. This had the effect of straightening the straw to a certain extent as it was pulled out. Next, it was watered with a bucket, or perhaps a watering can, to make it pliable. Thatchers said that if the straw was not damp enough, it would not 'lie'.

Next, the broaches are prepared. Cut from hazel some months ago while hedge-trimming or ditching, the lengths of split wood might have been soaked for several days in a trough or pond, again to make them more workable. The first thing the thatcher does is to 'wring' or twist them in the middle so that they can be easily bent

A demonstration of rick-thatching by the Devon war agricultural executive committee in 1944.

Well-made stacks at Addlestone, Surrey in September 1946.

into the shape of a fork without breaking. Too dry and they would snap. If the ends are not sufficiently sharp, he might give them an edge with his knife. These broaches will have to be stabbed cleanly into the stack to hold down the thatch, and the thatcher will want a swift, clean entry without a fight.

Between the broaches runs a string, if such was available. If not, then a rope was made by winding hay into tight bundles with a 'scud turner' which consisted of a bent iron rod with a hook at one end, and an offset handle in the middle. By one man turning while the other man fed hay, a grassy string known as a 'bond' emerged. The greater the team, the faster the thatching, and so to cover a twenty ton stack of hay in a day might require one man or boy pulling and watering straw, another couple engaged in making bonds, while another, senior man, stays aloft on the stack.

One by one, bundles of straightened straw, tied with a rope, are brought up the ladder to the thatcher. Starting at the eaves, he begins to lay them like roof tiles across the length of the stack before moving up to the next row, overlapping as he goes. To secure each bundle, which he pats and spreads till it pleases his eye, he passes a length of the rope secured either side of the sheaf by a broach firmly stuck into the stack. It looks a flimsy covering, but a good thatcher knows that no gale of wind he has ever experienced will move it. The stack is now safe, and will survive many a winter month.

In southern parts, summer may linger on through this month, but in the north it may still feel as if summer has never arrived. Where Atlantic winds bring frequent downpours, even in the supposedly dry spells, then the battle to win even some hay may still be raging. The chances here of seven drying days on the run are slender, and farmers don't have the luxury of merely letting their grass sit in the wilting sun until dry and ready to cart. They have to be cunning. Some draped their grass along wire fences which ran for miles, allowing as much of it to face whatever drying conditions might prevail; although sun may have been in short supply, good hay can just as well be made in a fresh, drying breeze. In fact, there is an old saying in the south which says, 'there is much bad hay made in a good year'. This means that in a hot summer, when you might think conditions are right for fast, trouble-free haymaking, then a crop which has been too quickly 'cooked' will not necessarily be a good one. That is some comfort for the poor souls with their crop strung out on wires. If they don't have it safely home very soon, it could all be easily lost.

The potatoes will now be at their best and demand to be lifted. No sooner is the back-breaking gleaning finished, but the punishing potato harvest is about to begin. This is the last time the entire family, women and children will be called to the fields before winter sets in and any luxuries the farmer's boy might be enjoying, such as education, will be put on hold. The potato crop *must* be gathered.

If ever a job required a few dry days, it was this one. Wet soil clings to potatoes for grim death and harvesting them from sodden soil takes almost as much earth from the field as it does potatoes. In a dry loam, however, the soil drops away from the crop leaving a glistening potato which can look like an unearthed jewel on a fine autumn afternoon.

Potato graip

Potato hand-basket

Assuming we have moved on from the times when a flat-bladed fork, known in Scotland as a 'potato-graip', was the only means of lifting the crop, this is a job for a pair of

Harvesting potatoes by hand in 1944 on a farm near Moretonhampstead on the fringe of Dartmoor.

horses. The potato plough is designed to run under the ridges in which the crop lies, lifting the soil as it goes, exposing the potatoes for pickers to handle. It has a blunt nose, to prevent damage, and as the soil runs up it, like over a plough, instead of being turned by a breast it passes over iron fingers through which the soil can drop, leaving the potatoes on top. This can be pleasant work for the horsemen, if going well, but a sharp-eyed farmer will be looking for cuts and damage and will have a few sharp words if he sees what he considers sloppy work.

The farmer will also be keeping a keen eye on the gatherers. If a large acreage was to be covered, then hired help might be needed and people could be employed from outside the district whom he did not know. If the pickers were paid by the weight of crop they lifted, the less scrupulous might be tempted to drop a hefty stone into their sack and make a little money the easy way.

The potato plough was the first of the mechanical aids to take the drudgery out of lifting this crop, but it still required the pickers to rake around the soil with their fingers to get the best of the yield. When the potato 'spinner' arrived, life became a little easier. This device, hauled by two horses, ran down the middle of the ridge, or baulk, and as it moved forward a revolving wheel, with fingers attached, flicked sideways through the ridge, flinging the potatoes aside. The beauty of this machine was that not only did it remove the potatoes from the soil completely; it dropped them in a neat row on clean ground. From here, they were loaded into baskets, and carried to a cart standing by the edge of the field. When this was full, it was hauled to the potato 'clamp' where they would be stored till ready for sale or winter feeding.

There was never a moment when you could say the crop had been entirely lifted, for the potato has a clever habit of hiding. No matter how much attention has been paid to the gathering, there will always be some that are left behind. After the plough, spinner and pickers, the drag rake takes to the field to remove the haulm, the dried

The potatoes which have been lifted by the Ransomes spinner on the left are being loaded into a tumbril to be transported for storage, probably in a clamp. Near Reading, Berkshire, 1934.

Village children were given a week off school to help with the wartime potato harvest.
Here a crop of Red King is being gathered in 1942 at Stillington in Yorkshire.

stalks, and this will no doubt reveal a missed potato or two, sometimes big ones. Next will come the harrows, and yet a few more will be uncovered. Any that are missed will make their presence felt later in the year when they sprout annoyingly amongst the following crop, which was usually wheat.

There was one incidental advantage in the work that the potato crop required. Throughout the summer, the soil has been covered by a thick overhang of foliage and stirred on a regular basis as the ridging plough went through the crop to earth up the ridges. It has provided a most difficult climate in which weeds might flourish, which was why potatoes left behind what was called 'clean' land. With no chemical weed killers, this was an important part of the rotation of crops around the fields of the farm.

Some might argue that at this point in the calendar, the farming year has come to

Rudimentary protection from the wind and wet, and a fire on which to brew tea,
was provided for these potato pickers at Hartlebury, Worcs, in 1942.

a close. It is true that with the exception of the root crops, everything that had been planned for this year has now been harvested, and thoughts turn to the coming winter and the plantings that need to be done then. The shepherd, too, if he has not yet put the ram to his ewes will be planning ahead. The horsemen will be thinking that it will soon be time to house their horses again for the winter, and get back into the old routines of dark mornings and evenings.

There are no beginnings and no endings on a farm. It is like being on a roundabout that never stops.

October

For some, October marked not only the end of a farming year, but the end of farming completely. Michaelmas Day, which falls in late September, was traditionally the day on which farms changed hands. October was the time of the farm auctions, when the accumulated possessions of a lifetime's farming came under the auctioneer's hammer.

Depending on the circumstances, these were either feasts or wakes. For a farmer retiring after a long and prosperous career, this was his send-off. But for a farmer bankrupted by the vagaries of the market, the weather, or just plain old farming bad luck, the farm sale was conducted under a cloud of gloom. No farmer likes to see another go under.

For all the relaxed atmosphere of a traditional farm auction, this was when farmers needed to be at their sharpest. If it was equipment they were looking for at bargain prices, then they might be suspicious of anything which had been given a fresh coat of paint, for what was the vendor trying to disguise? Perhaps it was no more than to cover the putty with which he'd filled the cracks in the wood; but what were the cracks doing there in the first place? Was the machine on its last legs, or had it been badly damaged and fit for no more than the scrapheap? Amidst the amiable banter, eyes were looking for all these tricks.

The most useful tool, on a farm auction day, was probably a pen-knife with a small but sharp, pointed blade. In a quiet moment, when no one was watching, a farmer might quickly stab at a wagon's wheel, or a seed drill's shaft, to see how soft they were beneath the paint, and try to estimate how far the woodworm had got. Was it worth buying cheap, and mending?

The traditional farmer took no holidays, had no idea of foreign travel, could find within his own parish all the company he ever needed, so farm sale days became

LOT SEVENTEEN. (Plan No. 4).

ALL THAT

FREEHOLD PROPERTY

KNOWN AS

COOMBE FARM, AXFORD

in the Parish of

RAMSBURY, WILTS.

The situation is in the Kennet Valley about midway between Ramsbury and Marlborough, South of the road and river, and within a very few minutes of the main London and Bath Road, where it passes through Savernake Forest.

The Homestead is placeed practically in the centre of the Farm with three short lengths of private roadway leading out on to the public roads.   The

LONG FRONTED BRICK-BUILT & THATCHED FARM HOUSE

contains :—Five Bedrooms, Two Staircases, Two Sitting Rooms, Kitchen with range, Dairy and Cellar.   Outside is a Brewhouse with Force Pump from Well to small Storage Tank over, Copper, Open Fire and Sink.   There is a Pump from Underground Tank near.

There is a Lawn in front and a Large Vegetable Garden in rear of the house.

THE FARM BUILDINGS

include :—**Brick, Stone and Thatched Range of Cart Horse Stable or Cowhouse, 3-Stall Nag Stable and Trap House** or loose box.

SUBSTANTIAL BRICK, FLINT & SLATED TWO MOW BARN

adjoining which are Two lean-to Cart Sheds.

Timber and Thatched Range of **Cowhouses for 17** with **4-Bay Cart Shed** in rear, **Thatched 2-Bay Implement Shed**, a Brick, Flint and Iron Roofed Open **Cowhouse for 15**, Two Timber and Slated **Loose Boxes or Calving Boxes**, Brick and Slate Range of **Five Pigstyes** and Fowl House.   There are

THREE BRICK AND THATCHED COTTAGES

at a little distance from the Farm Buildings.   Two of these are let off and one is void.

THE FARM LANDS, 120 ACRES (more or less)

ALL PASTURE, are divided into Six Convenient Paddocks, all of which have good boundaries. Trial Drinking Places have been dug in Two or Three of the Enclosures and could no doubt be further developed, especially that in **No. 702.**

There is an old Bridle track along the Northern side of **No. 689.**

SCHEDULE :—

NO.			DESCRIPTION.					AREA.
689	...	...	Pasture	...	...	...	...	35·127
699	...	...	Ditto	...	...	...	..	4·875
700	...	...	House and Building		...	...	...	2·265
701	...	...	Pasture	..	...	...	...	13·722
702	...	...	Ditto	...	...	..	...	34·066
702a	...	...	Ditto	...	..	...	...	10·269
702b	...	...	Ditto	...	...	...	...	9·870
723	...	...	Road	...	...	...	...	2·045
724	...	...	Pasture	...	...	...	...	8·257
727	...	...	Cottages and gardens		...	...	...	·340

A 120·836

Tithe :—Vicarial £8 3s. 8d.   Rectorial £36 8s. 10d.   Land Tax £1 10s.

The Purchaser will have to take to, by Valuation, all unconsumed Hay on the Farm at Spending Price.

13

Notice of a farm sale, 1927.

occasions for what were almost tribal gatherings. A farm sale worked towards a climax, the best and most highly prized being saved till last, and these were usually the cattle. Somehow, the cattle were the kings and queens of the farmyard and farmers were judged on the quality of their herds. You couldn't be a good farmer, yet have a poor herd - the two things simply didn't go hand in hand.

It was probably the sale of horses that brought lumps to throats, more so than with any other of the farm's livestock. After a long working life, a cart-horse became a true part of the family, its habits known as well as those of the children. There were no guarantees where they might end up when sold under the auctioneer's hammer, and the thought that they might find their way to a farm where they were less fed and cared for must have caused many a caring farmer a few sleepless nights. Of course, not all farm horses were 'good 'uns' and there was nothing like a farm sale for getting rid of a bad tempered colt, or a mare with a kicking habit. You had to keep your wits about you if you went buying at a farm sale.

October was also time to be thankful for the harvest just past. Parishes might hold their harvest festival at any time in this month, well after the harvest had been gathered in. September seemed such a busy month, and time flies.

The harvest festival was less a religious event, and more a feast. Held around the big table in the farmhouse kitchen, if the farm was large enough and sufficiently prosperous to support such an event, nothing was spared in feasting and toasting the harvest now safely gathered in. Writing as far back as the early sixteenth century, the poet Thomas Tusser compiled a long verse describing his view of the proper conduct of the farming year and he was in no doubt about the value of the harvest feast:

> *In harvest time, harvest-folk servants and all*
> *should make all together good cheer in the hall;*
> *And fill out the black bowl of blythe to their song,*
> *And let them be merry all harvest time long.*

Often, it was not the farmer himself who sat at the head of the table, but the Lord of the Harvest, the man who had brought together and taken charge of the harvesting gang. The farmer and his wife might well be the servants. And what a feast they would dish up. One writer spoke of, 'smoking puddings, plain and plum; piles of hot potatoes, cabbages, turnips, carrots, beef, roast and boiled, mutton veal and pork, everything good and substantial; a rich custard, and apple pies, to which the children did ample justice.'

And after the feast came the games and the singing, loud and long into the night, fuelled by the home-brewed beer which flowed in quantity as the harvest gang sang loudly of 'largess', which was a small sum of money, a tip, given to the men as they bundled the sheaves:

Now the ripened corn
in sheaves is borne,
And the loaded wain
brings home the grain,
The merry, merry reapers sing a bind,
And jocund shouts the happy harvest hind,
Hallo Large! Hallo Large! Hallo Largess!

But while we talk of feast and plenty, it is worth remembering that the less prosperous of the family farms would have no celebration at all, merely an unspoken thanksgiving for anything the harvest has provided. More fitting for them might be the grace,

O heavenly father bless us,
And keep us all alive;
There are ten of us for dinner
And food for only five.

The next day, with aching heads no doubt, they will resume their work, remembering the harvest feast until the next arrives to eclipse it.

The potatoes must be protected now for it will not be long before the autumn frosts begin to fall. This involved the building of a clamp, suitably placed on the edge of the field where the potatoes had grown, or nearer to the farmyard if the crop was likely to go into winter feeding. A flat piece of land was required, well drained where water wouldn't stand and cause the crop to rot from underneath. There seems to be nothing more infectious than the beastly disease which makes potatoes rot. It would be easy for just one rotten potato to infect several tons of otherwise healthy ones and reduce them to offensive, useless pulp. To make matters worse, the farmer might not have known the putrefaction was taking place until it was too late and he caught scent of the vile, spreading rot.

October 1941: tipping potatoes from a tumbril to build a clamp on David Black's farm at Bacton, Suffolk.

So, the base of the clamp is covered with dry straw to a good depth and the heap of potatoes built on top of that, as high as can be managed but making some effort to avoid using forks in case they cause damage. Then begins the covering, firstly with forked straw, which is allowed to settle. On top of that comes another layer until a good thickness is built up and it resembles a thatch. Ventilation was vital, so 'chimneys' were left in the crown of the roof, made by preserving a circular hole and filling it with loose straw to allow the air to flow but the frost to be excluded. Then, on top of all this was slung a good layer of earth. Properly made, a potato clamp could stand all winter and the crop takes no harm.

The mangels might be ready to harvest in October. These can be the mainstay of the winter feed on a livestock farm, and must be harvested with care. They are not lovers of frost, and although a mere ground frost will do them no harm, when the air frosts begin to bite, which can be in November, then the mangels should be out of

As soon as they were clamped, the potatoes were covered with good wheat or rye straw to protect them from winter frosts. These men were working near Rochester, Kent in October 1942.

the weather's reach. For all its appearance of being a bruiser of a crop, with its vivid, ruddy, swollen roots sometimes the size of footballs, it is, in fact, quite a sensitive creature. If it is cut it will 'bleed' to death. The chemistry of the mangel is fascinating. In the time from harvest to eating, which is some three or four months, the starches turn to sugars and so instead of something as 'woody' as a swede, the mangel tastes as sweet as a well-sugared bowl of fruit. It is little wonder that there is nothing on the farm, not even a chicken, which will shun a bit of mangel if offered.

But the harvesting must be careful and the root kept intact if the mangels aren't to bleed and rot while in store in a clamp, identical to the one built to store potatoes. The secret lies in the precise use of a knife to remove the leafy tops without slicing into the root itself. The leaves tend to be quite fulsome and will rot while in store leaving the mangels swilling in a green slime. They have to be removed. But to cut

Carting mangels at Allerford, Somerset in 1942.

*Tipping mangels (with their tops still on to save wartime labour)
from a tumbril to be clamped near Salisbury in 1941.*

Clamping sugar beet.

precisely, while lifting and at the same time throwing the mangel into a cart, is a great skill. Mastery of it makes mangel-lifting for an experienced hand, half the work it can be for a novice. Somehow, the upward movement of the root as it is pulled from the ground brings it past the blade of the knife held in the other hand. The upward, slicing movement of the knife then helps to propel the mangel upwards into the air and into the waiting cart. Mangel lifting is yet another of those skills which takes a second to perform, and often a lifetime to perfect.

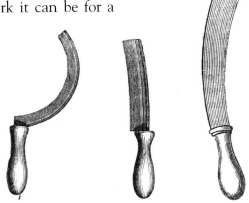

Turnip trimming knives

The last visitors to the corn stubbles, before the ploughs move in, are the pigs. On a traditional farm they did not feel the confinement of modern pig-keeping methods. True, the household pig spent much of its life in a pig sty near the farmhouse where scraps of otherwise waste food could be flung. But if the pigs were kept alongside other livestock, then there was hardly a place where they weren't allowed to forage for what others left behind. Their ability to remove every last grain of corn, even those left by the gleaners, made the pig an appealing animal to have around for while he was fattening himself he was also helping clear the land. An eye must be kept, though, for the pig knows that some feed can be won with far less effort and if it means barging a gate to get at some sweet clover on the other side, then the pig will get his mighty rump against that gate and have it off its hinges before the farmer has chance to chase after him with a stick.

It is to be hoped that any wandering pigs fail to find their way into the kale field, for now this green, leafy crop is beginning to look its best and the shepherd will be guarding it, knowing that this will be the basis of the feed for his flock when the meadows finally give up the struggle of growing and resign themselves to the coming winter. This is the last fresh greenery of the year.

For sport, a horseman or farmer proud of his ploughing skills might take himself off to what was known in the eastern parts of England as a 'furrow drawing match.' These were distinct from a ploughing match in that instead of a small parcel of land being turned over and judged, in the case of a furrow drawing match it was a single furrow which was ploughed and carefully measured. Of course, it made the stakes that much higher for in a conventional ploughing match, a little deviation in the work, a wobbly furrow or a poor start could be concealed with later ploughing, and marks lost early on could be won back later. In a furrow drawing contest, there was one shot only at it.

It was a true test of the working relationship between man and his pair of horses, calling for control and communication. The drawing matches were often conducted with just one pair of horses who during the course of a day would have to serve many masters, which didn't make it any easier for the competitors. It is one thing to draw a straight furrow with horses you know well, but working to such precision with strange horses propelled the contest into a different league. Here the difference between winning and losing might be as little as a deviation of a couple of inches over the length of a fifty-yard furrow - a mere wobble which would be of no matter in a twenty-acre field.

The horseman would bring with him sticks, painted white for visibility, to plant at

The Isle of Thanet, Kent, ploughing champion, 1953.

the far end of the furrow (if the rules allowed) whereas back on the farm he would have made do with a pale twig snapped from the hedgerow. A good horseman could draw to a single stick, having a natural sense of whether or not he was drawing a straight line. Others might use two sticks, one behind the other, keeping them in line as they drew. Whichever method they chose, without doubt the worst moment must have been when they reached the end of the furrow, turned to look back at their work, and saw how true a line they had drawn. Many must have wished the ground would open up and swallow them.

The judging of the furrow was conducted with great ceremony. A white stick was planted at each end of the furrow, and set vertically upright using a spirit level to guarantee precision. Another judge then works his way along the furrow till he thinks

High-cut work at the British ploughing championships, Grantham, 1958.

he finds the point of greatest 'deviation' and he plants a stick there. But, he may not have put it in the right spot, and there would be much debate and measurement until the worst bit of the furrow was found. Then, another stick is planted, this one in line with the two at each end of the furrow and shameful departure from the straight and narrow is measured. It was, apparently, not uncommon for no deviation to be detected.

It might be at this time of year, as the horses were once again put to the plough to turn stubbles into seedbeds for winter corn, that the slave of the farm, the farmer's boy, might glance out of the kitchen window to see the horsemen disappear to the fields and wonder how many more years it might be before he was one of them, in charge of his own horses instead of shackled to a scullery sink. The work of the fields was often unbearably hard, but the burden on the farmer's boy could be equally crushing. It was an endless round of chores from fire lighting at dawn, to pan scrubbing, floor washing, wood gathering. One remembered:

I used to have to get up at five o'clock in the morning, get fire going - and it was an old fashioned fireplace. Well, these brass pans, they were full of milk. One for the table and the other for the calves. Of course, a lump of soot fell down the chimney and into the pan, so that went to the calves and another had to be got for the men. This happened three times a day. Monday was wash day, Tuesday was baking day with all the washing up to do, Wednesday was floor scrubbing day, Thursday was cleaning the dairy, Friday was housework, and Saturday was worst of all - the master went off to market and left me to scrub the kitchen floors and feed the men at teatime. No stop, it was.

But there were no other choices: no career opportunities in rural communities for anything outside the farms and the businesses that supported them, such as the blacksmith, harness maker or miller. Everyone knew their place, and it was usually on the land. They learnt to be content there.

November

A winter month, or part of an extended autumn? It is always difficult to put a finger on November and decide whether it is friend or foe. If it should be a dry month, it might bring with it unwelcome frosts; if wet, then the last of the winter wheat, or beans, may be sown dangerously late. It was difficult for the traditional farmer to know what kind of a November to pray for; either way it could be a bleak time of year with spring seemingly an age away. As the old rhyme has it:

> No shade, no shine, no butterflies, no bees,
> No fruit, no flowers, no leaves, no birds,
> November.

Mornings might be foggy, making it a damp business when the shepherd has to move his hurdles in order to place his flock on clean grazing, or on kale. The sheep, fleeces sodden with dew and their own sweat, will brush against his legs leaving him damp for the rest of the day if he wasn't sharp enough to remember to protect himself. If old bones are going to ache in the coming winter, in November the first twinges will be felt. The early morning lighting of the farmhouse fire is now an urgent business rather than a luxury, and farmers' boys who have failed to secure enough dry kindling, or have not chopped sufficient wood, will carry fleas in their ears. In fact, the sound of saws could provide a background to much of this month, for although winter wood for the fires is best cut in the summer, when dry, the work of harvest, thatching and root- lifting meant less urgent jobs were left behind, and only now will the axes be swinging with any sense of real purpose.

For the horsemen, the early morning stable routine will set a pattern which will

Woodland work: Suffolk 1940.

last now till spring comes again. From the chaff-house will be heard the sound of the chaff cutter, slicing oat straw into short lengths for mixing with the horses' feed to fill their bellies and flesh out the more expensive part of the rations, like oats. At the back of the chaff house will be the heavy wooden chest, or 'ark' in which the feed is stored, carefully watched over by the head horseman or farmer to see that the horses get enough and no more. Of course, any horseman worth the name would ensure his horses got what he thought they needed, even if it meant stealing a little from the cattle ration, despite the farmer's orders. The secret was in not being caught. Stables were dusty

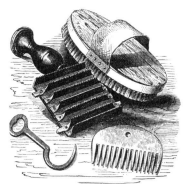

Curry combs for horses

J L Herbert milking by hand.

places, although fresh-smelling from having any droppings removed every day. But the spiders' webs which were strung from the rafters, dense with dust, were never removed, for horsemen knew that nothing irritated a horse more than an annoying fly and that there was no better fly trap than the cobwebs.

As the weather grew wetter, and the land became more sodden, part of the morning stable ritual was to tie up the horses' tails in intricate knots, fastened if necessary with thin strings of twisted straw. A well tied tail should hold its place from dawn till dusk. If a horse dropped its tail while at work, the man responsible would not be allowed to forget it.

The milking cows will have said their farewells to the pastures for another year, and in some farms will look forward to an entire winter in the byre. To stand next to a deep

Loading the milk churns at Oak Farm, St Albans, Herts, 1944.

littered cowshed in the winter is to stand next to a radiator, as not only the heat from the cows warms the air, but the rotting, trampled straw and muck is giving off yet more heat as it rots, emitting a not offensive smell - just the scent of cows in winter.

The management of the herd called for some concentration, depending on what use the farmer made of his milking cows. If he wanted fresh milk the year round to sell to the community or a local dairy then he must organize a staggered calving otherwise the supply, which only lasts for nine months or so, will dry up. Cheese makers on the other hand need plenty of quality milk which only comes from cows fed on summer pastures, and so these farms will calve in the spring and by Christmas the milk flow will have tailed off. But there are other factors in his calculations. Some of the calves might be for fattening for later sale, and these must be fed too. Nothing makes calves grow faster than their mother's milk and so the herdsman might find himself in the laborious position of having to milk the cow, keep half for the churn, and give half back to the calf, leaving the calf wondering why it wasn't allowed to do

Some aspects of traditional farming were not restricted to the countryside. David Carson's dairy at Cable Street, within a mile of St Paul's in London, kept going right through the Blitz. His milk van was setting off on its morning round in 1947.

it for itself. Greed, of course, is the answer for without doubt you would end up with a full calf and an empty churn. No self control, have calves.

On a traditional farm, the milking would be done by hand. In fact, hand-milking survived in England until the middle of the twentieth century. It had the obvious advantage that the machinery was never going to break down. Like riding a bicycle, those who can milk by hand are experts at doing it, but find it impossible to describe. There appeared to be two methods.

In the first, the teat is seized firmly near the root between the front of the thumb and the side of the fore-finger, the length of the teat lying along the other fingers. This pressure is maintained while passing them down the entire length of the teat. Two hands are used alternately, on different teats. Some would say this is not proper milking at all, and would properly be described as 'stripping.'

'Real' milking is a different action completely. Firstly, the whole teat is grasped with one hand, although the first finger and thumb grasp a little more tightly than the rest. Then, the milk is forced down the teat by progressively tightening the fingers, one after the other. Beneath the cow sits the bucket, to one side the milkmaid on her stool, and the only thing that comes between the farmer and his milk might be the cow's tendency to kick out, knocking milker and precious bucket of milk flying. The opportunity for bad temper is ever present although as Henry Stephens, in his *'Book of the Farm'* warns:

> *If a cow cannot be overcome by kindness, thumps will never make them better. And even when a young cow is cured of kicking, there is the ever present annoyance of the swishing tail to be dealt with!*

On a poor farm, where financial uncertainty was always threatening, as it was on many of the farms we are imagining, then the pig was the family's insurance policy. This is not the same pig that is part of a commercial herd, bred for fattening and butchering, but a pig that could feed the family.

The life of the backyard pig was a short but royal one. The pig lived in some comfort, well bedded and dry in a sty not far from the farmhouse door, and over the wall throughout the day came scraps from the kitchen table, green waste from the vegetable patch, a few potatoes from the bottom of a wagon, even a mangel or two if anyone felt kind-hearted enough. If a pig was really lucky, and the farm could afford it, then there was a daily feed of perhaps boiled barley, or 'middlings', which is one of

Feeding sows and their piglets.

the rough residues of milling, halfway in coarseness between the flour and the chaff. Sometimes, if the farm was near a brewery, the spent grains might find their way into the pig's ration. If near a bakery, a stale loaf or two. If the farm adjoined a village, then some of the residents might help feed the pig too, with peelings or surplus apples from orchards, in return for sausages or black pudding later in the year. Best of all was to be a pig on a dairy farm for then the skimmed milk and any surplus cream would be coming your way. On all these things, pigs would get fat. By which I mean *very* fat. No traditional farmer would recognize modern pig meat as being of any use for feeding a growing family or a working man. In the days of hard, continuous labour, the calorie requirement could be three times that of someone leading a sedate, modern lifestyle. Pig meat, with plenty of fat on it, could provide the energy.

But the pig had one other unique quality which made it a farmyard favourite, and why the month of November was a bad time of year to be a backyard pig. Of all the meats which were cheap protein providers in an otherwise poor diet, pig meat was the only one which could be preserved sufficiently to last an entire winter, and beyond.

There was nothing that could be done to beef or mutton which could preserve it in the way that pork could be cured into hams and bacon, which if properly kept could still be enjoyed a year after the pig had died.

The backyard pig, though, was a friend who had been greeted every morning, to whom troubles had been told, and whose back had been scratched repeatedly with a long stick. And now the pig must die. He was the family's insurance policy, because they knew that as long as they had a pig to kill in November, they would never starve whatever the winter might bring.

There was only a fleeting sadness of pig-killing day. It started early and the butcher, if one was employed, would come well before breakfast. Pans of boiling water were needed and the farmer's boy would have been up and about well before the sun, lighting fires, sharpening knives, feeding those on the farm who had work to do other than dealing with the pig.

The killing was swift and sure; a blow with a knife to the pig's jugular, somewhere just below the neck, a final twist of the blade to ensure it was well ruptured. From this gash the pig's blood would flow to be collected in bowls, prized by the lovers of black pudding. And with the flow of blood, the life of the pig ebbed away and it passed painlessly into oblivion. Now, the real work started.

The pig had first to be rid of its bristles. The buckets and bowls of hot water were brought from the kitchen, poured over the pig, which was then shaved top to tail, and scrubbed till the skin was hairless. By the time this was done, even a black pig was pink skinned.

How the butchering was done, varied. Some would work on the prostrate pig, cleaning away the innards before jointing the carcass. Sometimes it was hung from a beam, and carved. Whichever method was chosen, the imperative was to ensure no waste. The old saying is true - 'you can use every part of the pig but the squeal' and if you're wondering whatever possible use could be made of a pig's tail, then ask the children who made tormentors of them.

After the butchery came the processing. Before the blood had time to set, it was mixed with chunks of fat, oatmeal, salt and spices and poured into lengths of the pig's cleaned intestines. The hams, the sweetest and most precious part, were salted either by rubbing with dry salt for as long as three or four weeks, or immersing in brine, a salt solution. As the brine crept into the flesh and fat, it killed any bacteria in its path and this is why salted pork could be kept for months on end. Then, as Christmas approached, the hams might be moved from the cool larder or cellar to somewhere behind the chimney breast where not only the updraught of air would keep them 'sweet', but the smoke would add a whole new level of flavour to them.

By the end of pig-killing day, there would be very little left of the pig which hadn't been made into fresh joints of meat, hams, bacon, gammon or black pudding. All the left-overs, consisting of bits of liver, sweetbreads and the like, were gathered together, dropped into the pan, and the feast of 'pig's fry' brought to a close a vital day in the traditional farming year. The larder was now full to bursting.

No source of free food could be ignored. Throughout the autumn, fruits and berries would have been collected from trees and turned into jams and pickles; as yet another way of keeping their flavours alive till fresh food arrived in the spring. It was also a time to keep in check the numbers of rabbits on the farm, for as winter drew on and the rabbit became ever hungrier, then the kale, mangels, and newly sprouted winter wheat looked ever more appetizing. To some, rabbits were a crop. To a farm worker, a dressed rabbit or two could be sold to a butcher for a few pence, and although the farmer knew rabbits were a pest on his farm he still insisted that anyone removing rabbits from his land, other than himself, was poaching. Hares were a different matter - hares were classified as game and killing them was undoubtedly poaching.

Not all rabbit catching expeditions, however, had a satisfactory ending. In his *Farmer's Diary*, H. Rider Haggard remembers a farmer who convinced himself that his illness was without doubt heart failure. He went so far as to confirm the diagnosis:

> *he went with all precautions to London to interview a specialist, who, to his enormous relief, for he thought himself a doomed man, told him that his heart was perfectly sound. Investigations followed, and he discovered that his attack was brought on by eating cold rabbit pie by which he had been poisoned.*

If there are still winter crops to be sown, there is urgency now. December might be too late. If there is no time for ploughing, harrowing and drilling with the seed drill, the traditional farmer knew that the biblical practice of scattering the seed, or broadcasting, would produce some kind of a crop. This way, it was possible to plough and sow at the same time. On a piece of land which had been roughly ploughed, after a crop of roots such as turnips which the sheep have already devoured, the seed can be spread over the furrows and allowed to fall to the bottom of them. All it then takes is for a pass of the harrows and the seed is covered. There is also a chance that the crop will

Bagging up in the barn on a winter's day.

sprout in something resembling rows, roughly a furrow apart, where the seed has tumbled from the crown to the base of the furrow before being covered. It had none of the certainty of the seed drill, though, and it is doubtful if the rows would be sufficiently defined for the later use of horse-driven hoe to control the weeds between the rows. Nevertheless, it was better than leaving the ground bare as the year ebbed away.

Beans could certainly be sown in November, destined to crop the following year when they would be threshed and milled into rich feed for cattle, but used sparingly in horses for they had a ferocious reputation for 'heating' a horse.

The sowing of beans, and even wheat and barley, could be a tiresome process before the invention of the seed drill. A dibbler was used, which was no more than an iron rod about three feet long with a handle like a spade's. At the other end was a blunt, pointed end designed to be stabbed into the ground leaving a hole. Into these holes were dropped the seeds, one by one. To speed things up, the farm worker used one in each hand and walked backwards across the field making two rows of holes six inches apart. As with all such apparently simple, if tedious, jobs, there was more to it

Plough teams on their way home at Preston, Wilts. One of the many superb photographs taken by Eric Guy and now preserved at the Museum of English Rural Life, University of Reading.

than met the eye, and in the case of dibbling the secret was to give the dibbler a slight twist before removing it so as to leave a clean hole with firm sides which would not collapse before the sower came with the seed. To keep an even distance between the rows, and to stop them from growing ever further apart, sometimes the dibbling was done with crossed wrists, although the preferred method in Norfolk was to bring the feet together at every step backwards, and press the dibbler into the soil at the tips of the toes. The sowing of the seed itself was work for smaller hands, and so the women and children came to the fields. It was a slow way to sow a crop.

November comes to a close sometimes with gales and rain, but also calm, frost and fog. The milk yields have to be watched, for cows don't like cold nights. Working horses, returning from the fields hot and sweating, need careful attention now if they are to avoid chills. There is always a bit of lime-washing of stables and byres to be done if the weather allows no other job. Every day on a farm is too precious to be wasted.

December

In an uncanny echo of the distant days of summer when the song of the skylark was eclipsed only by the swish of the scythe, through the cold morning air comes the rasp of sharpening stone on steel blade. This time it is the hay knife which is receiving the stockman's attention.

The hay, of course, was not made into today's familiar bales; instead it was carted loose to the farmyard, built into a huge stack, and thatched for its own safety. Now, it is time to reap the rewards of all that effort. But hay demands a little more work before it finds its way into the stable, byre or stockyard. The cutting of it with a hay knife was hard work, as athletic as any job on the farm. The blade was broad and flat and would drag through the hay making cutting of it twice the work. Clearly, the sharper the blade the less the effort in cutting, but despite the honing many a farm worker who has climbed a stack feeling chilly has come down again dripping in sweat. Perhaps the hay knife was nature's way of keeping out the cold.

The hay, if properly cut, comes away in thick slices which are then forked onto wagons standing at the foot of the stack. From here they are hauled to cattle and sheep, every morning of the winter without fail, and again at the end of the day if necessary. By the time it arrived in the mangers, the hay might have travelled quite a distance since it was grass growing in a meadow. Cutting hay was part of the daily grind in winter, and the sharpening of the hay knife a prelude to every day's work as much as the lighting of the kitchen fire.

This was a hard-working time for the carts, wagons and tumbrils too. All the livestock were being fed as if in a

Hay
knife

Cattle on deep straw bedding: Wrotham, Kent, December 1933.

restaurant, all their meals brought to them, and so heavy loads of hay, mangels, kale, turnips would be endlessly on the move, wearing axles, making wheels shaky.

The wheelwright was a key figure in any farming community and without him there would be no one to mend shafts, fix bottom boards and axles, as well as the wheels themselves. None of his secrets were written down, and it is doubtful if he ever made much use of a measure. But his instincts were never far wrong, and the roundest of wheels he would make largely by eye, different numbers of spokes for carts and wagons, with some wheels as wide as six inches if they were destined to carry heavy loads without sinking into soft land. His tools were simple and effective, mostly chisels, saws, planes and spoke shaves. And with these he would cut hubs with mortices, shape spokes to the right length, cutting tenons into the ends of them where they would slot neatly into the 'fellies' which were the parts of the outer circle of wood on which the rim sat.

Wheels were rimmed in iron, which may have been made by the blacksmith if the wheelwright confined himself to working with wood. The 'hooping', or tyring, was a sought after job on cold winter days for it involved lighting fires, getting iron red hot,

Finishing a wooden wheel with a metal tyre, near Wantage, Berks, 1959.

running around, keeping warm. The iron hoop was first put on the ground and sticks of chopped oak were laid over it. Then, like a circular bonfire, the wood was set alight and the blaze fed till the iron rose to almost a red heat. With massive pincers, the strongest two men lifted the hot, expanded wheel rim and dropped it over the wooden wheel. They prayed it dropped easily. The wood, of course, would soon catch fire if left to its own devices, but buckets of water were now thrown over it, cooling the metal and contracting it at the same time till it became what the wheelwright hoped would be a perfect, tight fit. Then, once more the wheels of agriculture could start to turn.

No moment is wasted in December, for tradition holds that ploughing should be finished by Christmas. There is good reason for this. There is nothing better for turning lumpy clods of soil into crumbly earth, than a good frost or two. On a deeply cold night, the water in the clods will freeze, and then come the morning sun it will thaw and as it expands it will break the clods apart. Over successive nights it seems that even boulders might be reduced to a tilth by this process. In fact, all it takes is a pass or two of the harrows in the spring and ploughed land turns into a perfect

December 1942: ploughing at Broadwood Hall, Herefordshire.

seedbed. No winter frost is to be wasted for it does its work effectively, without complaint, and most importantly at no cost.

Only one thing brings all the men from the fields, and that is the arrival of the threshing machine. On threshing days, all hands are needed and only the essential feeding and milking continues without interruption. Threshing time is one of the great set pieces of the farming year and is when the farmer finally learns the success or otherwise of his farming efforts, for although he may have been well pleased with the number of sheaves he carted from the corn fields, he has no way, until now, of knowing how much wheat or barley they contained.

Before the invention of the engine-driven threshing machine, which could devour an entire stack in a day, the process of removing the grain from the sheaves was long, physical and brutal. It employed a flail, which is made in two parts. One is called the 'hand staff' and is a piece of ash four or five feet long. The other half is known as the beater, or swingel, and is shorter and stouter and might be made of a tougher wood such as holly or blackthorn. The two pieces of wood were capped with either a bent piece of ash, or possibly ram's horn, to form a loop, and between these a piece of knotted leather or eel skin.

In 1949 Tom Hancock, a smallholder at Brook, Kent,
was still using a 150-year-old flail for threshing peas and beans.

Portable threshing machine   Clayton and Shuttleworth's finishing threshing machine

It was truly laborious work carried out on the floor of a barn with the barn doors open to allow the breeze to blow through to carry the chaff and dust away. The hand staff is grasped firmly with both hands, raised above the head, and swung in such a way that the swingel almost completes a circle of the labourer's head before being brought down with some force onto the ears of corn on the floor. It is like beating the corn into submission, and made all the more difficult as two men usually worked together, carefully matching their swings to avoid hitting each other. Imagine the delight at the invention of the steam driven thrashing machine. Or perhaps not; there is an old saying in Essex, according to Thomas Hennell, writing in *The Old Farm*, which says:

> *the sound of the threshing would tell one how the work was being paid;*
> *a slow, dull thumping announcing 'by the day, by the day', while*
> *piece-work would induce a brisker rhythm: 'we-took-it, we-took it,*
> *we-took-it'!*

Of course, the work was far from over once the threshing was complete because what remained on the threshing floor was a mixture of corn, chaff, straw and dust; all the hard work had so far achieved was to beat the grain from the seed head. Winnowing was done by the breeze, which might have been created by hand if a winnowing machine was used, or simply provided by the good Lord and made available by opening the barn doors on a windy day. The crudest method was simply to wait till a fine breeze blew through the barn, then take spades and shovels and fling the threshing residue high into the air. With luck, the lighter dust and chaff would travel further on the breeze than the heavier corn, and the two would come to rest in two heaps. The chaff, of course, was not entirely wasted, for it could be used to bulk up horse and cattle feed. The winnowing machine did a similar thing, but the

A 1943 threshing scene at Sutton, near Ely, Cambs. In the centre the 50-year-old steam engine is driving the threshing drum. The men are feeding the crop from the depleted stack in front. The threshed straw is being elevated to the stack at the right and the sacks of corn are being loaded on to the cart at the left.

draught came from a hand-driven fan and the falling corn dropped through sieves leaving a cleaner sample of corn.

The threshing machine did all these tasks in one, and left corn, chaff and straw in three distinct places - corn in sacks, chaff in a pile, straw in a stack. Only a large farm could afford one, small-holdings would rely on the services of a threshing contractor whose arrival would be heralded by the rumble of his steam engine as it made its stately progress down the lanes, the threshing machine in tow. Sometimes, if the engine was not self-propelled, the entire outfit would be hauled by a team of horses. If a chaff cutter was being employed to cut the straw, then a small army of horses and horsemen were needed to haul the engine (four horses), the threshing machine (four horses), the elevator (two horses) and the chaff cutter (two horses).

Arrival at the farm was no certainty of anything; members of old threshing gangs will tell countless tales of farm arrivals where, after rain, the farmyard would be flooded, the ground soft, and the heavy machine sinking slowly up to its axles. But somehow or other the stacks had to be thrashed for there was another farmer anxiously waiting his turn, in need of the money which those sacks of corn represent.

The engine had a prodigious thirst, and maintaining a constant supply of water was a job for the farmer's boy. He would usually set up a trough somewhere near the engine, keeping it filled with buckets if there was no pipe near to hand. Running out of

Wheat being fed into the threshing drum at Worlingham, Suffolk, 1944.

water was the worst crime he could commit on threshing day for the whole operation would grind to a halt as the engine ran out of steam, and everyone would know exactly who was to blame.

It could take as long as a day to set up a complete threshing machine and engine. Most important was that the farmer chose to build his stacks in the right place to start with. Too close together and it was difficult to place the tackle between them. Too far apart and it made harder work for the men who would have to pitch the sheaves across to them.

It was the farmer's responsibility to provide the coal, hard Welsh coal being the preferred fuel. A good day's threshing would use as much as a quarter of a ton. First, the 'drum' as the threshing machine was known, had to be set exactly level, employing a spirit level for precision. The body of the machine was made of wood, although the mechanism was iron, and all manner of vibrations ran through it subjecting it to all sorts of loads from all directions. A machine which did not sit true would not run for long without a major failure.

Then the engine had to be placed in position, exactly in line with the drum for a long

*The dirty end of the job: at the rear of the drum this chaffing attachment
was preparing animal feed in 1936.*

belt was to be passed between the two, and tightened. If the drum and engine were not in precise line, the belt would spin off before the first revolution had been completed. The driving of the engine, the running and maintaining of it was a huge skill, and on those farms where no tractor or engine driven device had ever been seen before, the skills of the engineer must have seemed to belong to someone born on another planet. The engine driver learned to spread the coal about on the grate, and not simply shovel it into a pile thus preventing a good draught. He had to understand the workings of the dampers to produce just the right amount of heat without wasting coal. He had to know where to apply the oil, when to let off steam. And he had to remember that one of the first jobs of the morning was to put a few potatoes in the smoke box at the front of the engine so that they would be nicely baked come dinnertime!

Threshing employed more men on a farm than any other job. It was difficult to do the job properly with less than a dozen. Most important, after the engine driver, was the 'feeder'. He balanced himself on the top of the swaying threshing machine and grabbed the sheaves as the pitcher on the stack flung them to him off the end of his fork. In one

A busy threshing scene in December 1936 not long before the years when the push for greater agricultural production jeopardised many of the good practices of traditional farming.

hand was a knife and as the sheaf landed in his grasp, he gave a quick flick with the blade through the string that was holding the sheaf together. Now he fed it into the machine, not simply dropping it like a bunch of flowers, for too much at once and the machine would falter, which was when driving belts flew off and tempers frayed. Instead, he let it fall off his hands in a controlled and even way so the machine laboured steadily instead of in fits and starts. There was some danger in the early machines: the man whose job was to feed the thrasher, stood over the open jaws from where he could easily fall in, to a certain death. Such threshing accidents were not unknown.

Once the machine had the sheaf within its workings, it could start the process of beating it against the walls of a revolving drum (hence the nickname of the machine). Through the walls of the drum, the chaff and corn could pass, but the straw couldn't. The straw, instead, was carried to the far end of the machine from where it fell out of a gaping mouth. An elevator was usually positioned here which lifted it into the air, dropping it where a couple of men with pitchforks built it into a stack. The corn and chaff travelled in a different direction, blown by a fan. The chaff, dust, rubbish, weed seed and general foulness appeared, as if blown by a gale of wind, out of one side of the machine, and the precious grain appeared at spouts on the other side, each spout delivering grain of a different size. At the grain spouts stood the farmer, assessing his crop, deciding if it added or detracted from his financial prospects. At the chaff spout, up to his neck in filth, stood the farmer's boy. Each to their place.

After the departure of the threshing machine, a peace fell over the farm. The routines of feeding and watering continued. On frosty mornings, the shepherd would be breaking the ice on the troughs to allow the sheep to drink. Horsemen would inspect the fields to see if the frozen land would yield to the plough. There was a sense now of having come full circle, and all the worries and anxieties that had been eclipsed by the summer months start to surface once again. The shepherd will be fretting about his ewes, for lambing will be only a few weeks away. He prepares his hut for the long nights he will spend with his flock. The farmer sees that already the daily slices taken from the haystack are beginning to show and, using only his eye and his experience, he will hazard a guess as to whether he has enough put by to see him through to spring. The mangel clamp will remain undisturbed for a few more weeks, till the kale has been finished, but the mangels will be all the sweeter for waiting - providing the frost makes no inroads. Cattle that have been slow to fatten, which he fears may show him little profit, will head to market now. And if he has kept a few farmyard geese or turkeys, he will remember that Christmas is not far away and the butcher might pay well for them.

Despite the shortness of the days, the ploughs are at their busiest. Horsemen will walk their horses to the field, often in darkness; the sparks from their shoes as they make their way along the flinty lanes providing the only light. And there they will both stand till dawn finally comes, when one end of the furrow can again be seen from the other, and the day's ploughing will commence, eleven miles to be walked to plough an acre.

And will, as tradition demands, the farmer be able to place the breast of his plough under his bed on Christmas Eve as a mark that not only has his ploughing been complete, but as an offering to whoever holds the fortunes of farming in their gift?

It's doubtful. Did ever a farming year run according to plan?

About the Author

Paul Heiney

For over a decade, Paul Heiney immersed himself in the practices of traditional farming on his 40 acre holding in East Suffolk. Having learnt to work Suffolk Punches under the brief but intense teaching of Roger Clark and Cheryl Grover, he was determined to try it for himself and keep a farming tradition alive. His adventures were recounted to the wide readership of *The Times* through his weekly column.

He is a writer and well-known broadcaster working on a wide range of programmes, but most recently his recreations of the Victorian Farming Spring, Summer and Winter have drawn large and appreciative audiences for Anglia Television.

Paul is married to writer and broadcaster Libby Purves.

Other Titles from Old Pond Publishing

A Victorian Summer *Anglia Television*

In his first series of programmes Paul Heiney captured the hearts of East Anglians with his thoughtful re-creations of working lives in the Victorian countryside. Video.

A Victorian Winter and Spring
Anglia Television

Paul Heiney heads back to the Victorian farm to re-enact two of the toughest seasons of bygone farming. Video.

The Implements of Agriculture
J A Ransome

Originally published in 1843, this generously illustrated book summarised the developments of ploughs, thrashers, drills, milling machines and a wide range of horse-powered implements. Hardback.

Harnessed to the Plough
Paul Heiney with Roger and Cheryl Clark

A year on a farm run with Percherons and Suffolks. Everyone who loves horses and traditional farming skills will be enchanted with this programme. Video.

Joseph and His Brethren
H W Freeman

First published in 1928, this novel follows the story of a Suffolk farming family through two generations. Paperback.

Chaffinch's *H W Freeman*

H W Freeman's moving novel depicts the life of farm worker Joss Elvin and his struggle to raise a family on 19 acres of Suffolk farmland. Paperback.

Early to Rise *Hugh Barrett*

A classic of rural literature, this is a truthful account of a young man working as a farm pupil in Suffolk in the 1930s. Paperback.

A Good Living *Hugh Barrett*

Following on from Early to Rise, Hugh takes us back to the assortment of farms with which he was involved from 1937 to 1949. Paperback.

In a Long Day *David Kindred and Roger Smith*

Two hundred captioned photographs of farm work and village life in Suffolk 1925-33. Paperback.

For further details and a full list please contact
Old Pond Publishing
Dencora Business Centre
36 White House Road
Ipswich
IP1 5LT
United Kingdom
Tel: 01473 238200 • Fax: 01473 238201
Website: www.oldpond.com